TAXONOMY AND BEHAVIORAL SCIENCE

Comparative Performance of Grouping Methods

This is a volume of

Quantitative Studies in Social Relations

Consulting Editor: Peter H. Rossi, University of Massachusetts, Amherst, Massachusetts

A complete list of titles in this series appears at the end of this volume.

TAXONOMY AND BEHAVIORAL SCIENCE

Comparative Performance of Grouping Methods

JUAN E. MEZZICH
University of Pittsburgh School of Medicine
Pittsburgh, Pennsylvania 15261, USA

and

HERBERT SOLOMON
Department of Statistics
Stanford University, California 94350, USA

1980

ACADEMIC PRESS
A Subsidiary of Harcourt Brace Jovanovich, Publishers
London New York Toronto Sydney San Francisco

ACADEMIC PRESS INC. (LONDON) LTD.
24/28 Oval Road
London NW1

United States Edition published by
ACADEMIC PRESS INC.
111 Fifth Avenue
New York, New York 10003

British Library Cataloguing in Publication Data
Mezzich, Juan E
Taxonomy and behavioral science.—(Quantitative studies in social relations).
1. Cluster analysis 2. Social sciences—Statistical methods 3. Psychology—Statistical methods
I. Title II. Solomon, Herbert III. Series
300'.1'51953 HA31.3 80-40750

ISBN 0-12-493340-8

Printed in Great Britain by
THE LAVENHAM PRESS LIMITED
Lavenham, Suffolk

To Enrique, Eudocia, Julio Cesar, Ada and Ada Maria Mezzich

To Lottie, Mark, Naomi and Jed Solomon

Contents

List of Tables

Preface

In the early 1970s Dr Juan Mezzich, trained as a psychiatrist, was engaged in clinical diagnosis and in this way became interested in quantitative methods for the classification of individuals into groups. This prompted his formal graduate study at Ohio State University for the PhD in mathematical methods in psychology. In 1974, while on the faculty of the School of Medicine at Stanford, his work on his dissertation required formal supervision, which led to a meeting with Dr Herbert Solomon of the Statistics Department who had been suggested by colleagues at Ohio State. From this initial interaction both worked very closely on classification and clustering techniques. This subject had been of interest to Dr Solomon for some time and so the partnership became a very active one, with many meetings over a two-year period. These meetings led not only to the completion of the dissertation but also to this exposition.

Multivariate data analysis has had a tremendous resurgence in recent years, sparked by ingenious developments in computer hardware and programing technology. Much of what began in grouping methods by investigators in the late nineteenth and early twentieth centuries was blocked by the sheer computing effort involved and led to a 50-year period of development of mathematical models and analytical techniques in multivariate studies. In this volume we steer a completely orthogonal course, dealing only with data-dependent analysis. In this way, exploratory analysis of the data is accomplished by techniques that provide groupings of individuals or elements for investigators in the behavioral sciences as well as in other disciplines.

This book offers a non-mathematical description of clustering methods before entering into the empirical comparative study. Some knowledge of matrix algebra may be useful when reading sections of Chapters 1 and 2, but this level of mathematics is certainly not required in order to read and employ the results developed in the book. More extensive descriptions and mathematical developments of clustering can be found in some recent books such as "Mathematical Taxonomy" by Jardine and Sibson and "Clustering Algorithms" by Hartigan, which are listed in the References.

Four data bases are supplied to serve as a basis from which a number of different clustering procedures appearing in the literature can be examined. Except for the iris data, the others were developed for this study since data bases do not abound in the literature and can usually be found only in the private files of individual investigators. The procedures are discussed in Chapter 2 and the data bases in Chapters 3, 4, 5 and 6. These procedures are then analyzed and evaluated in Chapters 7 and 8, and in the last chapter one is applied to a rather large data base of socio-economic data on household expenditures. The hope is that many investigators in diverse areas, finding themselves with large data bases, can profit from the results given in this volume. Since the procedures are completely data-dependent, a large variety of topics can yield to the findings.

We would like to offer our appreciation and thanks to several US Government agencies, under whose aegis some of the manuscript was developed. They are the Office of Naval Research, the Army Research Office and the Veterans Administration. Our thanks go also to Mrs Carolyn L. Knutsen for her skill, patience and graciousness in the preparation of this manuscript for the publisher. Some others should be mentioned for their help: Professor James Erickson for advice and suggestions on issues of statistical and theoretical psychology; Professor Helena Kraemer for general statistical advice; Mr Dwayne Ball for inestimable assistance in data processing and software development; our colleagues in the Department of Psychiatry at Stanford who participated in the development of archetypal psychiatric patients; and our wives Ada Mezzich and Lottie Solomon for their general support and encouragement.

January 1980 *J.E.M.*
H.S.

1

The Quantitative Taxonomic Approach in Behavioral Science

1.1 INTRODUCTION

There has always been a need to achieve parsimonious yet operationally meaningful accounts of what is going on in nature and in human behavior. It appears that the human being is by nature a classifying animal as his functioning and survival seem to have depended on his ability to recognize and communicate similarities and differences between objects and events in his universe. The development of the human mind has been closely related to the perception of discontinuities in nature (Raven, Berlin and Breedlove, 1971). Ontologically, it has been proposed that the development of reason in the child is fundamentally involved with 'grouping' operations and dichotomous distinctions (Piaget, 1973). These observations point to the dual nature of all classification: on the one hand, the urge to name, number and sort all that has impressed man as different, and on the other hand, to search for order among that which appears to be amorphous (Temkin, 1965).

The meaning and role of classification in human life in general are of central importance in science. This is so because classification involves two basic scientific functions: (i) the description of objects of interest or under investigation, and (ii) the establishment of general laws or theories by means of which particular events may be explained and predicted (Hempel, 1965). These functions can be attributed by extension to taxonomy, which is the science of classification. Furthermore, a taxonomy can also represent a model of reality and, as such, can embody or reflect a theory about how a particular domain is structured and how it works (Fabrega, 1976). For example, it has been proposed that in naming and explaining an area of cultural relevance, taxonomies provide a way for members to relate and behave in a culturally appropriate manner (Kay, 1971). Within medical taxonomy, diagnosis is the focal point of thought in the treatment of a patient: backwards to pathogenesis and etiology, forwards to prognosis and treatment. Consequently, diagnostic categories provide the locations where clinicians store the observations of clinical experience and the diagnostic taxonomy establishes the patterns according to which clinicians observe, think, remember and act (Feinstein, 1967).

We are aware of attempts by biologists to classify flora and fauna, although even that dichotomy was a major step forward. It is in the physical and life sciences that we find the first quantifiers at work on such matters. Later, we find social anthropologists and psychologists engaging in studies by which groups can be created. Today, we find taxonomy pervasive in practically every field of study.

Even though we regard classification in social sciences as rather new, it is difficult to think of its counterpart in physical sciences as very old unless one thinks of a few hundred years in the course of mankind as a very long step. It was just two or three hundred years ago that many ailments were labeled 'consumption', because they were characterized by a 'wasting away of the tissues'. Under this were lumped such diseases as leprosy, tuberculosis, diabetes and others. It was not until some time later that someone noted that the urine of some of these sufferers was sweet and that of others was not. Of course, the subsequent discoveries of two different microorganisms for leprosy and tuberculosis suggested finer groupings that obviously were more meaningful in connection with specific treatments.

In addition to chronological differences in their taxonomic development, there are significant contrasts between taxonomic elements and purposes in botany and zoology and those in the medical and social sciences. In biological taxonomy, the cases ('operational taxonomic units', as labeled by Sneath and Sokal, 1973) are individual organisms considered with their total life spans and which are classified to reflect and enhance our understanding of morphology and phylogenetic evolution. In the medical sciences and in psychopathology, the 'operational taxonomic units' are particular human beings at a given time or during a given time span in their life histories who are classified to facilitate professional communication, prognosis, treatment selection, etiological understanding and public health measures. Special taxonomic issues here include the variability in the manifestations of a disease during its course, the possible coexistence of various morbid conditions, and the problematic reliability or reproducibility of meaningful clinical judgments and observations. In the social sciences, the taxonomic elements are quite variable including, for example, individuals in studies of personality types, human populations in some anthropological studies, and social structures and institutions in sociology and political science. The purpose of such investigations are predominantly to summarize and organize data and to elucidate data patterns. At later investigational stages, the purposes range from social intervention to theory development. Particularly problematic taxonomic issues here are the reliability and validity of the variables assessed to describe the entities to be classified.

It is worthwhile to emphasize at this point the exploratory nature of most taxonomic work in the behavioral and social sciences. Accordingly, analytic methods used in these areas have included not only formal grouping or clustering methods but also methods such as ordinal multidimensional scaling (developed within the framework of psychological research) which allow visual and more candid appraisal of data patterns. This focus on exploratory data analysis is quite appropriate in behavioral science, given the early state of its scientific development. As Bronowski (quoted by Feinstein, 1967, p. 72) warns, 'A science which orders its thought too early is stifled', because a science 'cannot develop a system of ordering its observations . . . until . . . [it] has passed through a long stage of observation and trial'. The focus and thrust of this exposition are on exploratory data analysis and the following chapters should demonstrate this.

1.2 HISTORY

There is evidence of classificatory efforts since the dawn of mankind. Egyptian descriptions of physical and behavioral disorders attributed to the physician Imhotep date back to 3000 BC. Ayurvedic literature from the last centuries of the Pre-Christian era include the *Caraka-Samhita* and the *Susruta-Samhita*, which contain classifications of medical illnesses including mental disorders conceptualized as demonic possessions. Hippocrates (460-377 BC) considered mental disorders, including epilepsy, as derived from natural rather than supernatural factors and classified them according to such variables as chronicity and presence or absence of fever. Temkin (1965) emphasizes that the origins of the classification of illnesses go back into a remote past, and that the beginning of the medical sciences lies somewhere along the road, not at its start.

Folk taxonomic systems have been found from very ancient times and are distributed widely throughout the world. As indicated by Raven, Berlin and Breedlove (1971), recognition is given to naturally occurring groupings of organisms in all languages. These groupings appear to be treated as psycho logically discontinuous units in nature and are easily perceptible. In most folk taxonomies, groupings that are members of the level *generic* (Bartlett,

1940) are finite in number and in fact are remarkably similar in size, ranging from 250 to 800 forms for either plants or animals. Reflecting on the essential role of classificatory activities in life, the study of folk taxonomic systems provides an opportunity for elucidating human cognitive processes as well as for understanding the usefulness of taxonomic systems themselves. Early biologists, such as Theophrastus (372-287 BC) quite directly reported on folk taxonomic systems prevalent in those early days (Raven, Berlin and Breedlove, 1971). It appears that folk taxonomic systems are essentially appropriate for communication with those who already know about the nature of the entities involved. They are not designed for information retrieval and the absence or presence of this feature seems to be a key differentiator between folk or traditional and 'modern' taxonomies.

A major era in the history of taxonomy started in the eighteenth century with the Swedish biologist Carolus Linnaeus and his work on a comprehensive taxonomy of plants and animals, for which he used a binomial system. He gave tremendous impulse to the idea of a systematic examination of nature in search for stable order and patterns. Inspirationally, he said in his *Genera Plantarum* (1737): 'All the real knowledge which we possess depends on methods by which we distinguish the similar from the dissimilar . . . We ought therefore by attentive and diligent observation to determine the limits of the genera, since they cannot be determined *a priori.* This is the great work, the important labour, for should the genera be confused, all would be confusion.' His example reinforced the teachings of his contemporary, Sydenham, on stable disease entities with fixed manifestations, and encouraged the development by Boissier de Sauvages and Cullen of monumental nosologies composed of thousands of species of disease, organized into classes, orders and genera.

The statistical stage in classification was ushered by the beginning of anthropometry. A pioneer here was Camper (1791) with his work on measurement of facial angles. In psychiatry, around the onset of the nineteenth century, Pinel and Esquirol inaugurated the formal study of mental statistics and emphasized clinical observation, reacting against the specific disease entity tradition of Sydenham. Quetelet, in 1835, was the first to present frequency distributions of human anatomical measurements.

In the late nineteenth century we find a blossoming of quantitative inquiries into classification through the selection and appropriate use of manifest variables. Quite often a one-dimensional index that incorporates all pertinent variables was sought so that a technician could assign an individual to one of several groups based on his reponses to the variables employed. For example, the coefficient of racial likeness was an index developed at the turn of the century to distinguish different national or tribal groups on the basis of a set of physical measurements. Inquiries on association of criminal types with physical measurements of individuals also received attention in this period by such investigators as Lombroso. Much of this inquiry took place in the British community of scholars and, in a way, it might be viewed to have begun at least in a larger sense with Charles Darwin's collection of data arising from his travels around the world. His diaries presented many observations on the animal kingdom and served as a base for study by many who came later in the nineteenth century.

It was with these investigators in the last quarter of the nineteenth century that we have the beginnings of statistical contributions to classification. In fact, it is the classification problem that in a way motivated and created statistical inference as an area of scientific inquiry. The modern discipline of statistics was brought about by the anthropometrists, biologists and psychologists of that era. Such initial contributors to modern statistics as Francis Galton and Karl Pearson stem from that period.

Galton seemed to be perpetually engaged in data analysis. He and his cousin, Darwin, and others matured in an age of scientific inquiry that emphasized empiricism. Pearson, along with others, later attempted quantification and mathematical treatment of the empirical analyses provided by their colleagues. Galton, whom we regard as the founder of regression analysis through his study of relationships between the heights of children and their parents, also initiated and developed the notion of correlation prior to 1885. The correlation coefficient serves as a basic summarization in multivariate data analysis and consequently in studies that go into techniques of grouping. From its very nature, a high correlation coefficient would indicate that the two variables belong in a group and a low correlation would suggest that they do not.

In one of his papers in 1888, Galton became interested in the classification problem. He pointed out that 12 measures proposed by Bertillon to be used for classification of fingerprints were not independent and suggested that the observed measurements be transformed into a set of independent measures. He also suggested the method of transformation, which we can now view as simple or unweighted summation in factor analysis. Thus, at quite an early stage we see the intermingling of classification analysis and factor analysis which, of course, is still quite current.

Pearson was engaged in studies that were obviously related to classification. In an interesting paper in 1901, he discussed mathematical representations of lines and planes of closest fit to systems of points in space. This geometrical way of looking at the classification problem may present a neater view of the problem to some. In effect, the multi-dimensional observations at hand, e.g. six measurements (age, IQ, schooling, history of drug use, crime of violence, length of stay in prison) for each member of a prison population of, say, 500 members can be viewed as 500 points in a six-dimensional space. Moreover, each point cannot be reached by traveling along six perpendicular axes, for the six variables can and usually do have degrees of association which must be taken into account. Only if they were independent of each other would orthogonal axes serve a useful purpose.

This problem—of finding a grid of fewer orthogonal axes to replace the grid of correlated axes (naturally, the points remain where they are)—is fundamental to multivariate data analysis. If the number of dimensions can be reduced to two or three without too much loss of information, some ease is achieved since elements can be grouped by eye. It is in this context that factor analysis is now employed in numerical taxonomy.

The works of both Galton and Pearson anticipated the development of factor analysis, but it was later that two psychologists brought this to fruition by their conceptualizing and measuring of human intelligence. First, Spearman (1927) formulated a two-factor theory at the turn of the twentieth century, and then Thurstone (1938, 1947) developed multiple factor analysis by building on Spearman's model in a strictly mathematical way.

Also during the first half of the present century, a number of interesting taxonomic efforts in various behavioral and social sciences took place. In linguistics, Kroeber and Dixon (1903) provided an areal classification of Californian languages based on grammatical pattern. In 1928, Czekanowski (cited by Driver, 1965) applied a proximity coefficient to the interrelationships among nine Indo-European languages and produced an intercorrelation matrix with the magnitude of the entries represented by four shades from white to black to facilitate visual appraisal of patterns.

In ethnology and social anthropology, pioneering work was done by Boas and Kroeber around the early part of this century. For example, proximity coefficients were used first to determine geographical clustering of ethnic units and then to reconstruct the history of these ethnic units. Czekanowski (1911) provided an interrelationship of 17 traits of material culture among 47 African tribes and clustered the material into two geographical- historical units.

In archaeology, early seriation studies were conducted by Kidder (1915), Kroeber (1916) and Spier (1917). Strong (1925) developed an intercorrelation of four time levels in Ancon, Peru across 40 descriptive traits of representative pottery, and showed that the four time levels could be telescoped into two major levels.

Classification in psychology, since the pioneering work by Spearman and Thurstone on factor analysis, was until relatively recently dominated by this technique. In 1935, Stephenson proposed the use of Q-correlation coefficients computed between individual across variables, which were useful for developing typologies through Q-factor analysis. Zubin (1938) described an interesting clustering procedure for subdividing a group into subgroups of like-structured individuals and for determining the factors that make them like-structured. A relatively recent application of Q-factor analysis was the grouping of tribal inventories explored by Driver and Schuessler (1957). Even more recently, innovative clustering methodology related to Q-factor analysis was developed by Lorr and Radhakrishnan (1967) and by Skinner, Reed and Jackson (1976).

During the past two decades there has been a tremendous expansion in the development of methods related to pattern recognition and clustering. In fact, a fully fledged field of quantitative taxonomy has developed, including complete clustering techniques (both hierarchical and partition types) and intermediate methods involving representation of multivariate data units. Several representative methods are reviewed in the first two sections of Chapter 2.

Some grouping-relevant techniques, such as ordinal multidimensional scaling (Shepard, 1962a, b; Kruskal, 1964a, b), were originally developed under the framework of behavioral science. Regardless of the field or purpose for which they were originally developed, most clustering techniques have been applied to the exploration of patterns in a large and growing number of areas of behavioral science.

1.3 ASSIGNMENT PROCEDURES AND DISCRIMINANT ANALYSIS

It is important to be specific about the term ‘classification’. This assumes that the data have already been clustered into groups and the assignment of data to these specified groups is desired. Actually, the latter process can be viewed as a subset of the former. On many occasions, we require data to produce both the number of groupings or clusters and the assignment of each element or individual to these groupings. In the assignment problem, the number of groups or clusters is predetermined. Each group is labeled, and rules are designed on the basis of which an assignment of each element is made to one of the fixed groups.

We do not wish to convey a sharp distinction between clustering and assignment procedures. If a classification procedure is not producing meaningful groups through the assignments that are made then changes are called for, namely, revising the predetermined groupings either in number or in shape or in both on the basis of the new information. This sequential revision of groups on the basis of the data available at different times suggests that one is indirectly engaging in clustering procedures. On the other hand, it is wise to keep in mind the conceptual differences just mentioned between attempts at clustering and attempts at assignment.

An essential step in classification or clustering procedures is the representation of the relationships among the variables or the entities, whatever the case may be, on which data have been collected. Among other important and prior steps, there are the processes of developing numbers to measure phenomena, making decisions on the employment of nominal, ordinal or continuous data, and subsequent coding of these data for analysis. We do not review these issues here, but we are mindful of their impact on the data analysis that will undergo investigation. Thus, we return quickly to clustering and assignment techniques and the basic summarizations of data for these purposes.

The clustering and assignment problems, even though they were recognized for some time, were not formally handled until the 1930s. The assignment problem received the first impetus. The analysis was provided by R. A. Fisher—one of the great savants of modern statistical inference, who was also a principal contributor to genetics. In a paper in 1936, we find what is now Fisher's classic work on discriminant analysis: 'The use of multiple measurements in taxonomic problems', published in the *Annals of Eugenics.* The author was to say, somewhat later, that the paper was written to embody the working of a practical numerical example arising in plant taxonomy in which the concept of a discriminant function seems to be of immediate service. This is a simple but fascinating statement, because it demonstrates once again that it is only when there is a problem requiring solution that some strides can be made. Too often we find solutions looking for a problem, and this is something we should be especially concerned with in classification and clustering problems.

In his paper, Fisher also listed the basic data he analyzed. This is rarely done by authors, and we therefore find the Fisher data and only a few other data bases referred to time and time again by subsequent authors who are experimenting with new assignment or clustering techniques. In this way, an anchor is provided against which the results of other techniques can be assessed.

The data employed by Fisher were supplied by a botanist and represented measurements on the irises of the Gaspe Peninsula. These data were previously published in the *Bulletin of the American Iris Society* and was therefore not a likely contender for a bestseller. Since it is a classical piece in the statistical literature, let us look at it in some detail. Four measurements on each of 50 plants in each of three iris categories were obtained. The categories are: *Iris virginica, Iris versicolor* and *Iris setosa.* For each of the 150 plants already assigned to one of three categories, there are measurements of sepal length, sepal width, petal length and petal width.

If we refer back to our geometrical representation, we have 150 points scattered in a four-dimensional space, except that each point is already labeled as belonging to one of three groups. The question is whether in some neat and simple way we can separate the 50 points belonging to any one group from the other two sets. This is compounded by the fact, in this case, that two of the iris species, namely *I. versicolor* and *I. virginica*, actually have a specific genetic relationship and do have some overlap in their morphology. In other words, Fisher is looking for hyperplanes that partition the four-dimensional space and that, after partitioning, will hopefully leave each group inviolate. Algebraically, he is asking for a linear function of the four measurements (later called the discriminant function) that accomplishes this. As a reasonable index for determining the coefficients of the linear function, he suggests one that will maximize the ratio of the difference between the means to the standard deviations within species. At times in this text, and these are very few, we resort to some mathematical jargon. However, throughout the monograph we are constantly concerned with quantitative thinking, and hope the English language exposition is simple and incisive. To be specific in this case, let d_p, $p = 1,2,3,4$ represent the difference in the observed means. Then for any linear function, X, of the measurements, namely,

$$X = \lambda_1 x_1 + \lambda_2 x_2 + \lambda_3 x_3 + \lambda_4 x_4$$

the difference between the means of X in the two species is

$$D = \lambda_1 d_1 + \lambda_2 d_2 + \lambda_3 d_3 + \lambda_4 d_4$$

while the variance of X within species is proportional to

$$S = \sum_{p=1}^{4} \sum_{q=1}^{4} \lambda_p \lambda_q S_{pq}$$

where S_{pq} is the sum of squares or products in X_p and X_q.

The particular linear function that best discriminates the two species will be one for which the ratio D^2/S is greatest, by variation of the four coefficients λ_1, λ_2, λ_3, λ_4. Geometrically, we are locating the hyperplane that best separates two groups of points in the sense that the distance between the four-dimensional centroids is greatest. Even though there are three groups of irises, in effect Fisher acts as if there were two groups, since *I. versicolor* and *I. virginica* are genetically tied together. Note that the variations within species is assumed to be the same in this development.

The index that is employed to provide the delineation is tied at first to the multivariate normal structure assumed for each species. Yet it is very similar to the indices suggested by non-parametric multivariate data analysis, as we will see in the next section. Here we are maximizing the difference between the centroids of the two species of irises or, in other words, maximizing heterogeneity between groups. This theme will be carried through all of our attempts at classification and clustering: we will either maximize heterogeneity between groups or minimize the scatter (i.e. seek homogeneity) within groups.

As a result of the analysis, Fisher arrives at a linear discriminant function that accomplishes a good separation. For example, *I. setosa* is separated completely from *I. versicolor* and *I. virginica.* It turns out that only one of the four measurements is really necessary to do this, namely petal length, and this can probably be seen by simply looking at the 150 sets of measurements. This should be something for us to highlight, especially when we discuss data sets for which meanings are not so specific and measurements are not so commensurate. This will obviously be so in any number of studies in behavioral science.

Fisher's work has been extended to assign an element to any one of *k* groups, and computer programs exist in computer center libraries to accomplish multiple linear discriminant analysis. Attached to this subject is the question of how many variables should be used in a discriminant function. It is clear that the more variables one uses, the better the discrimination should be, but it is also clear that the marginal gain in using additional variables can decrease sharply and therefore some variables can best be omitted in the interests of parsimony. Thus, we seek the best discriminating variables.

We might also ask what one would do if one were faced with the 150 irises and did not know their groupings; that is, if we had only the four measurements on each, and we wished to see what number of groupings and assignments could be made; this is developed in Chapter 5. Here, we are no longer faced with the assignment problem alone but with the clustering problem or grouping problem which, of course, subsumes an assignment problem. It is to this topic that we now turn.

1.4 DATA SUMMARIZATION

It is useful, in talking about groupings, to consider whether we are grouping measurement variables or grouping individuals or entities of a population. For the iris data, we are grouping entities of a population. Quite often one is interested in grouping measurement or test variables, and this is one of the queries that is addressed by factor analysis. The basic data summarization in multivariate data analysis will depend on whether we are grouping variables or entities. We will resolve this in subsequent discussion by first examining in some detail the question of data summarization. However, in our analyses we sometimes employ correlation coefficients where distance measures are more appropriate because many investigators use these; it is then possible to contrast the results with other procedures, as is done in our later chapters.

There are several ways to begin the data summarization, all give a picture of data interrelationship but each has special advantages for any particular investigation. One representation is that displayed by n individuals or entities each measured on p variables (n points in a p-dimensional space) by a matrix with p rows and p columns where an element in the i-th row and j-th column, say t_{ij}, is the sum of the n cross-products of measurements (taken around the mean) on variable x_i with measurements (taken around the mean) on variable x_j. In brief,

$$t_{ij} = \sum_{r=1}^{n} \sum_{r'=1}^{n} (x_{ir'} - \bar{x}_i)(x_{jr} - \bar{x}_j),\ t_{ij} = t_{ji}, \bar{x}_i = (\sum_{r=1}^{n} x_{ir})/n.$$

Let us label this matrix $\mathbf{T}$. Naturally, an element in the main diagonal, say the i-th row and the i-th column, is the sum of the squares of the deviations of x_i from its mean. If $p = 1$, then $\mathbf{T}$ is a scalar, namely

$$\sum_{r=1}^{n} (x_r - C)^2 \text{ where } C = \sum_{r=1}^{n} x_r/n.$$

If each element in the scatter matrix $\mathbf{T}$ is divided by n, the resulting matrix is the covariance matrix with cell entries s_{ij} and we label this $\mathbf{K}$. Now if we also divide each element, s_{ij}, in $\mathbf{K}$ by the standard deviations of x_i and x_j, the resulting element $r_{ij} = s_{ij}/s_i s_j$ is the correlation coefficient between x_i and x_j and the resulting matrix is now the correlation matrix which we label $\mathbf{R}$.

An important advantage of $\mathbf{T}$ is the manner in which it can be decomposed into two matrices that are especially pertinent in clustering and classification studies. In a classification study, the n entities will be assigned to k predetermined groups. The i-th group with, say, n_i entities can be viewed as a universe with its own scatter matrix formed as before and labeled $\mathbf{W}_i$. If we sum all the $\mathbf{W}_i$ scatter matrices, we obtain $\mathbf{W} = \sum_{i=1}^{k} \mathbf{W}_i$ and let this represent the 'within' scatter or homogeneity of the groupings. Likewise, for each of the k groups, we compute the group mean (a p-dimensional vector where the r-th coordinate is the mean value based on the n_r observations for x_r) and then produce the $(p \times p)$ matrix that we label $\mathbf{B}$, for it expresses a measure of the 'betweenness' or heterogeneity of the k

groups. The central point in this development is the existence of the fundamental matrix equation

$$\mathbf{T} = \mathbf{W} + \mathbf{B}.$$

This result suggests immediately an index by which classification (a predetermined number of groups) can be evaluated and, by extension, how clustering can be terminated at some cluster size. For any given data set, **T** is fixed. Thus, measures of 'groupiness' or 'clusteriness' as functions of **W** and **B** are thrust forth for examination.

For $p=1$, the matrix equation reduces to an equation about scalars. Thus, a good grouping index is one which minimizes W or, equivalently, maximizes B. We may also consider maximizing either the ratio B/W or $T/W = 1 + B/W$. An added benefit is that this ratio is invariant under linear transformations of the data. Statisticians have long exploited this fact, for B/W multiplied by an appropriate constant is the familiar F ratio in the analysis of variance.

When the number of measurements per element is two or more ($p > 1$), grouping criteria are not so straightforward. Several possibilities suggest themselves and have been developed and studied by investigators. One criterion suggested by several authors, that is a quite natural index, is the minimization of the trace of **W** (sum of all elements in the main diagonal of the matrix) over all possible partitions into k groups. This is equivalent to maximizing trace **B** because

$$\text{trace}\,\mathbf{T} = \text{trace}\,\mathbf{W} + \text{trace}\,\mathbf{B}\,.$$

However, trace **W** is invariant only under an orthogonal transformation and not under non-singular linear transformations.

Another criterion that may be employed for $p > 1$ is the ratio of the determinants

$$|\mathbf{T}|/|\mathbf{W}| = |1 + \mathbf{W}^{-1}\mathbf{B}|\,.$$

We can use $|\mathbf{T}|/|\mathbf{W}|$ as a criterion for grouping and select that grouping for which this index is maximized or, equivalently, $|\mathbf{W}|$ is minimized. Also we may employ $\log(|\mathbf{T}|/|\mathbf{W}|)$ since it is a monotonic function.

Another criterion for grouping is the trace of $\mathbf{W}^{-1}\mathbf{B}$ and we select the grouping that maximizes this index; it has been used as a test statistic in multivariate statistical analysis, as has the ratio $|\mathbf{W}|/|\mathbf{T}|$. The latter was employed by Wilks to test whether groups differ in mean values, and the former has been put forth by Hotelling in some situations and by Rao as a generalization of the Mahalanobis distance between two groups for $k > 2$ groups. We will shortly define and discuss the implications and uses of the Mahalanobis distance in clustering procedures.

Both trace ($\mathbf{W}^{-1}\mathbf{B}$) and $|\mathbf{T}|/|\mathbf{W}|$ may be expressed in terms of the eigenvalues, λ_i, of the matrix $\mathbf{W}^{-1}\mathbf{B}$. We write

$$|\mathbf{T}|/|\mathbf{W}| = \prod_{i=1}^{p} (1 + \lambda_i)$$

and

$$\text{trace}\, \mathbf{W}^{-1}\mathbf{B} = \sum_{i=1}^{p} \lambda_i$$

where λ_i are the roots of the polynomial formed from the determinantal equation, $|\mathbf{B}-\lambda\mathbf{W}| = 0$. The characterization of these ratios in terms of eigenvalues is helpful in data representation, especially when the effects of some reduction in dimensionality is desired. All the eigenvalues of this equation are invariant under non-singular linear transformations of the data; thus, measurements can be changed by scale or translation factors without affecting grouping results. It can be proved that these eigenvalues are the only invariants of **W** and **B** under non-singular linear transformations.

1.5 DISTANCE MATRIX

Thus far we have discussed some summarizations of multivariate data in matrix form, either **T** (scatter), or **K** (covariance) or **R** (correlation), and the kinds of grouping criteria that are suggested by the **T** format. Intuitively, we see that any grouping criterion is a function of homogeneity within groups, and heterogeneity between groups and the indices already described are specific quantities embodying these notions. We shall discuss other indices as we proceed, but each will be affected by the degree of homogeneity within groups and heterogeneity between groups. Attempts will be made to minimize the former, to maximize the latter, or in effect to do both. For the correlation coefficient index, large values indicate homogeneity while small values indicate heterogeneity.

Another method of summarizing data that is more appropriate on occasion is to find the distance between each pair of the n points in the p-dimensional space. This leads to a representation in matrix form of an $n \times n$ matrix where each element, in the i-th row and the j-th column, say d_{ij}, is the distance in the p-dimensional space between the i-th entity or individual and the j-th entity or individual. All the elements in the main diagonal are zero. The distance matrix is akin to the correlation matrix in that both may be viewed as similarity matrices—the jumping-off place for clustering attempts.

The correlation matrix is the natural beginning point in factor analysis where parsimony in the number of latent measurement variables is a desired goal. Historically, the main aim is to reproduce the correlation matrix with many fewer variables by examining the eigenvalues of the correlation matrix and retaining only those whose ratio with the sum of all eigenvalues is larger than the reciprocal of the order of the matrix. We will return to factor analysis and its place in clustering in subsequent sections. In some taxonomic situations the question of which measure of similarity to employ, whether it is of the association or distance type, will require some thought. While we will touch on these points, these inquiries will not be features in this exposition.

The notion of a distance matrix will be placed in sharper focus and this will be done by some discussion of appropriate distance measures. Because we will normally think of our data bases for clustering individuals or entities as n points in a p-dimensional space, the distance measures usually appropriate and available are Euclidean distance, Mahalanobis distance and city-block distance. The Euclidean distance between individuals or entities with respect to all p measurement variables may be written in vector notation as

$$d_{ij}^2 = (P_i—P_j)'\,(P_i—P_j)$$

where d_{ij} is the Euclidean distance between individual i and individual j, P_i and P_j are column vectors each with p rows listing the p measurements on the i-th and j-th individuals, respectively. The product of the difference row vector $(P_i—P_j)'$ by its transpose is a scalar. This is the distance function with which most of us are familiar. The city-block distance is given by $|d_{ij}| = \sum_{k=1}^{p} |p_{ik}—P_{jk}|$, i.e. the sum of the absolute differences of the coordinates of the two points.

The Mahalanobis distance may be written as in the notation above as

$${}_{m}d_{ij}^2 = (P_i—P_j)'\mathbf{W}^{-1}(P_i—P_j)$$

where $\mathbf{W}^{-1}$ is the inverse matrix of $\mathbf{W} = \sum_{i=1}^{k} \mathbf{W}_i$ and $\mathbf{W}_i$ is obtained for each of the $i = 1,3,\ldots,k$ groups by

$$\mathbf{W}_i = \sum_{m=1}^{p} (P_{mi}—C_i)\,(P_{mi}—C_i)' .$$

Note that a grouping of entities (which is the problem to be solved in cluster analysis) is necessary to compute $\mathbf{W}_i$ and consequently $\mathbf{W}$. Thus, the Mahalanobis distance takes into account the associations or interrelationships in the measurement variables. If two measurement variables are highly correlated, the Euclidean distance can be misleading because of the equal weight it imposes inaccurately on each measurement variable but this will not be so with the Mahalanobis distance. The Mahalanobis distance is more tedious to compute and for a long time it was avoided for this reason alone, but the computer has now made this computation more feasible.

Let us now look at some of these concepts in specific detail in some simple instances. For example, we have discussed the total scatter matrix **T**, the within scatter matrix **W**, and the between scatter matrix **B**. From **T** we can develop the covariance matrix **K** and the correlation matrix **R**. Consider the one-dimensional situation and the five points 0, 1, 6, 7 and 11 where, of course, the mean is 5. Assume two clusters: (0,1) and (6,7,11). The two groups have means respectively equal to $\frac{1}{2}$ and 8. Note how we write **T**, **W** and **B** in Fig. 1.1.

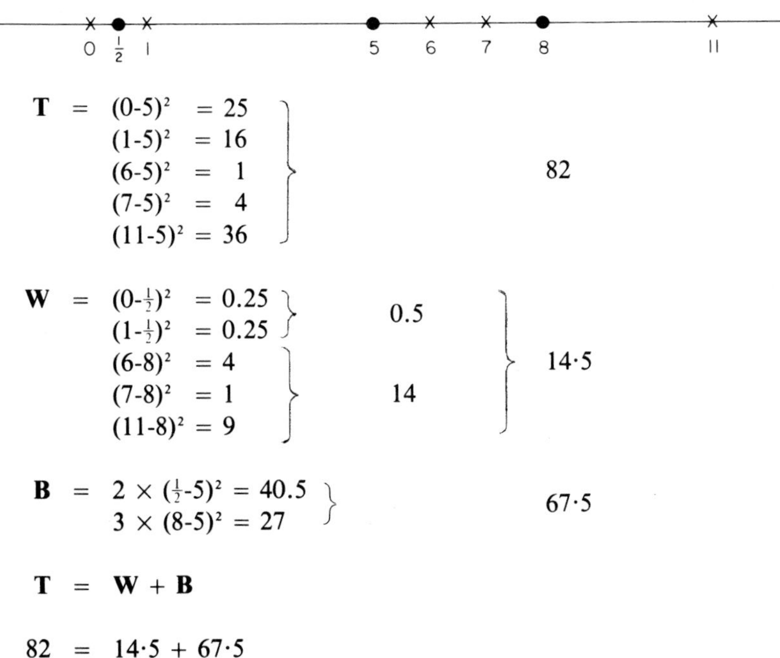

This, of course, extends to higher dimensions.

Fig. 1.1.

Suppose we now view the two-dimensional situation. Let us view the two-dimensional data matrix **J** with five pairs of observations. From this we compute the 5 × 2 matrix $\mathbf{D}_M$ where each entry is the entry in **J** minus its mean value. The 2 × 2 scatter matrix **T** is now obtained by the matrix multiplication $\mathbf{D}'_M\mathbf{D}_M$ as shown below. From this easily follows the co-variance matrix **K** which, of course, is 2 × 2 just as the scatter matrix **T**. We then normalize **T** to obtain the correlation matrix **R** as shown below. This is accomplished by dividing each entry in the i-th row and the j-th column by $\sqrt{(k_{ii}k_{jj})}$ (see Fig. 1.2).

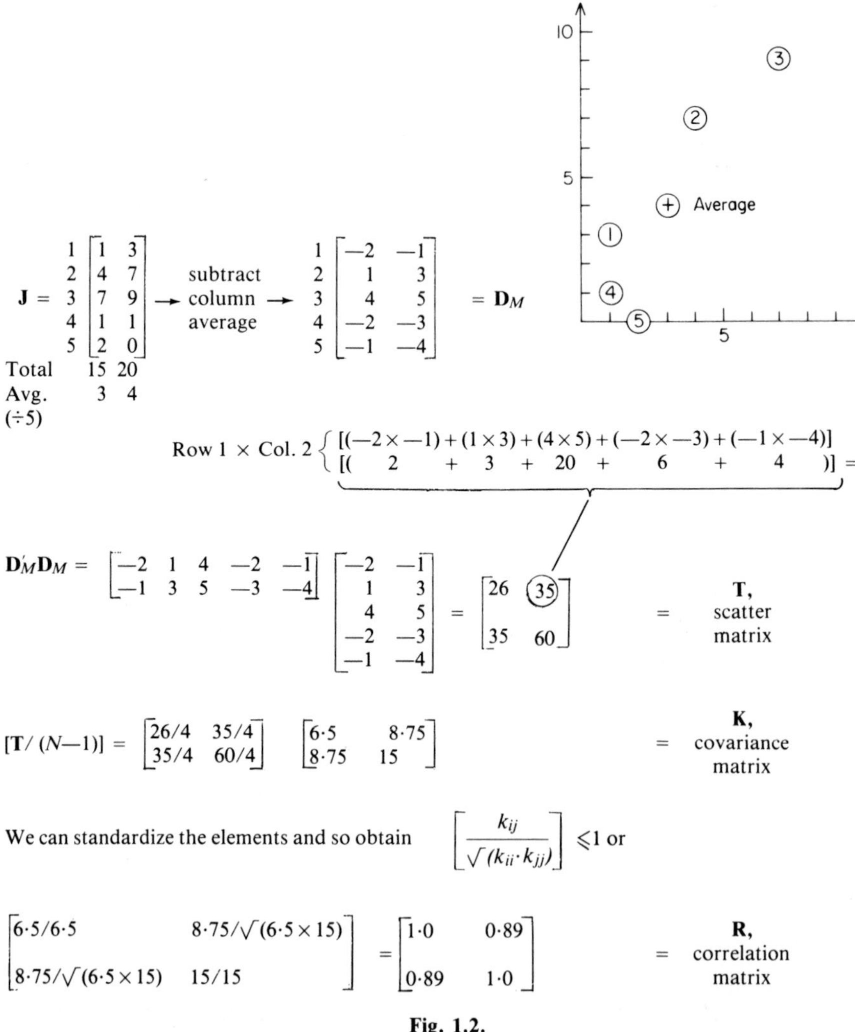

10
5
10
5
Average
J =
1 2 3 4 5
1 3
4 7
7 9
1 1
2 0
Total 15 20
Avg. 3 4
(÷5)
subtract column average
1 2 3 4 5
—2 —1
1 3
4 5
—2 —3
—1 —4
= D_M
Row 1 × Col. 2 [(—2×—1)+(1×3)+(4×5)+(—2×—3)+(—1×—4)]
[(2 + 3 + 20 + 6 + 4)] = 35
D′_M D_M =
—2 1 4 —2 —1
—1 3 5 —3 —4
=
26 35
35 60
= T, scatter matrix
[T/(N—1)] =
26/4 35/4
35/4 60/4
6·5 8·75
8·75 15
= K, covariance matrix
We can standardize the elements and so obtain
k_ij / √(k_ii·k_jj) ⩽1 or
6·5/6·5 8·75/√(6·5×15)
8·75/√(6·5×15) 15/15
=
1·0 0·89
0·89 1·0
= R, correlation matrix

Fig. 1.2.

The covariance matrix could have been obtained by dividing each element of the scatter matrix by N rather than by N—1 and thus obtain

$$\mathbf{K}^* = \begin{bmatrix} 5{\cdot}2 & 7{\cdot}0 \\ 7{\cdot}0 & 12 \end{bmatrix}$$

and consequently

$$\mathbf{R}^* = \begin{bmatrix} 5{\cdot}2/5{\cdot}2 & 7{\cdot}0/\sqrt{(5{\cdot}2\times 12)} \\ 7{\cdot}0/\sqrt{(5{\cdot}2\times 12)} & 12/12 \end{bmatrix} = \begin{bmatrix} 1{\cdot}0 & 0{\cdot}89 \\ 0{\cdot}89 & 1{\cdot}0 \end{bmatrix}$$

essentially the same as before.

The Friedman-Rubin clustering algorithm we describe in the next chapter stems directly from the matrix equation $\mathbf{T} = \mathbf{W}+\mathbf{B}$. However, the notions of internal homogeneity and external heterogeneity are appropriate indices by which to weight the effects of clustering rules.

In the preceding paragraphs we have discussed the various distance measures that are used by investigators in clustering. Other measures of similarity or relationships are association measures and correlation measures. The association measure is one that is used more frequently by investigators in biological taxonomy. There, variables are usually classified as 1 or 0, depending on whether or not a characteristic is present. A measure of association then, is,

$$\chi_{xy} = \frac{n_{xy}}{n_{xx}+n_{yy}—n_{xy}}$$

where n_{ij} is the number of ones that occur in the vectors x and y over the same variable. This index will not be pertinent to the kinds of measures of similarity we desire, and so we will no longer refer to it in this exposition.

Correlations, on the other hand, will be quite important to us. The most common correlation measure is the product moment correlation coefficient, or, as it is sometimes called, the Pearson correlation coefficient. This is defined in the following way,

$$r = s_{xy}/s_x s_y$$

where s_{xy} is the covariance between the vectors that express the profiles of x and y, and s_x and s_y are the respective standard deviations. Other versions which are not product-moment correlation coefficients, for example the tetrachoric correlations, are also employed by investigators who desire an index of relationship.

Experience has shown that the actual choice of the measure of correlation employed does not affect clustering algorithms too much. With this in mind one might just as well use the ordinary product-moment correlation coefficient, since program packages for either large- or small-scale computers are designed to compute this very rapidly. Note, again, that we are just employing correlation coefficients as descriptive indices and we do not concern ourselves with sampling questions or with the distribution of values of correlation coefficients.

Once the similarity matrix is developed, either by using a correlation measure or a distance measure, the investigator is primed to forge ahead with clustering procedures. We now turn to a discussion and development of a number of clustering procedures. Once these are produced it will be necessary to evaluate them and in the subsequent chapter we go into descriptions of clustering algorithms and methods for evaluating the efficacy of these clustering procedures. It will turn out that we have to return to measures of association in our evaluation—but we are jumping slightly ahead of our story.

Perhaps we can end this chapter with an account from real data. 25 characteristics on each of 82 aphasic children were collected at the Stanford Medical Center. The Mahalanobis distances between each pair of aphasic children were computed and each became an element of the 82 $\times$ 82 distance matrix that could now be formed. Prior to this analysis, each child had been classified by clinicians into one of seven categories for which treatement would then begin. On the basis of one of the clustering algorithms (King centroid procedure) discussed in the next chapter, most of the 82 children seemed to fall into one of two major clusters—thus casting some doubt on the meaningfulness of seven clusters. One of the two clusters contained children with either hearing or neurological impairment or maturational lag, and the other children with either mental retardation or autism. This made operational sense to the clinicians involved as being more in line with suggested treatment.

2

Clustering Procedures and Design of an Empirical Evaluation of Quantitative Taxonomic Methods

2.1 INTRODUCTION

The main emphasis here is on data-based clustering. We wish the data to tell us what is going on. *It is not so important that an exact number of clusters with just the right elements in each cluster (an ideal never achieved) is determined.* It is of value to partition the n multidimensional points in a p-dimensional space (measurements on p variables) in a valid, reliable and parsimonious manner, and in an efficient way.

However, any attempt at direct clustering or even classification of multivariate data awaited the advent of the computer as we now know it. In an earlier period we find a number of mathematically elegant multivariate analysis models provided by the ingenious work of scholars like Fisher, Hotelling, Wilks, Rao, Anderson and a number of others who directly or indirectly tried their hand at multivariate analysis. Fisher's multiple linear discriminant function technique (originally designed for two populations) to which we have previously referred, is now a classic and is a part of many statistical computer packages. As the reader will see, a principal contribution of this volume is the portrayal of four data bases and the assessment of a number of clustering procedures and measures of similarity by applying them to each of the four data bases given in Chapters 3, 4, 5 and 6.

The number and variety of quantitative taxonomic methods developed to date are such that they constitute now a rather wild forest, as recent surveys have shown (Ball, 1970; Spence and Taylor, 1970; Cormack, 1971; Sneath and Sokal, 1973). A need for some order in this forest seems to be obvious if a systematic appreciation of their worth is to be attempted. A classification of quantitative taxonomic methods which seems to be reasonable for the purpose at hand is outlined here.

I *Representation of multivariate data units (unfinished clustering)*

This category includes methods useful in taxonomic work but which do not produce a finished cluster configuration. Therefore, additional work, usually from a human being, is required to obtain a finished set of clusters with specified membership. Most of these methods were not developed for taxonomic purposes, but rather to represent variables in a multidimensional space. However, they can be used at least as major parts of a taxonomic process, and some of them (e.g. Q-factor analysis) have been heavily employed in behavioral taxonomy (e.g. Overall and Klett, 1972).

(A) *Methods using a relationship matrix (ordination).* This group of methods is characterized both by the use of a matrix of relationship measures as a necessary step in the taxonomic process and by its goal to place all entities in an orthogonal multidimensional space. Generally, two- or three-dimensional ordination is used for convenience of inspection and representation. Examples of methods in this group are Q-factor analysis and ordinal multidimensional scaling (Shepard, 1962a, b).

(B) *Methods not using a relationship matrix.* This class includes methods of subjective clustering from a graphical representation of multidimensional data. Therefore, these methods are quantitative only in their first part (namely, data representation) requiring as second part the use of the configural ability of the human mind in order to achieve cluster definition. The best known of these methods seems to be the faces developed by Chernoff (1973).

II *Finished cluster analysis*

This group includes most of the quantitative methods with specific taxonomic purpose. These methods characteristically produce a finished cluster configuration in the sense that their outcomes are sets of clusters with defined membership. Most of these methods can work with several types of relationship matrices. Therefore, their differences lie primarily in the second part of the taxonomic process, namely cluster determination. Cluster analysis procedures can be divided into two major categories: hierarchical and non-hierarchical.

(A) *Hierarchical methods.* These produce a tree-like taxonomic system, in which at one end, every entity is a cluster and at the other end all entities are included in a common cluster. The hierarchical cluster configuration is usually represented by a dendrogram. Besides constituting taxonomic configurations which are very appealing to some groups of scientists such as botanists and zoologists, hierarchical classification could also be seen as a family of partitions, each with a different number of clusters. In this sense they can also be appealing to researchers such as social scientists, who may be particularly interested in different partitions of the total group of entities or individuals. Hierarchical methods can be divided into agglomerative and divisive.

(1) *Agglomerative methods.* These methods construct the hierarchical tree from branches to a single root. They are among the most frequently used taxonomic methods. Specific agglomerative methods differ according to their definition of distance between an entity and a cluster or between two clusters. The single linkage method defines the distance between two clusters as that between their closest members, the complete linkage method as the distance between their farthest members, and the average linkage method as some type of average distance between clusters (frequent ways of defining average distances is through arithmetic averages and centroids).

(2) *Divisive methods.* These methods construct a hierarchical tree by beginning at the root and working towards the branches. They include monothetic and polythetic methods according to the number of variables used for each splitting of the data set. Divisive methods seem to be rarely used.

(B) *Non-hierarchical methods.* Non-hierarchical methods, in contrast with hierarchical ones, produce configurations that do not present rankings in which lower-order clusters become members of larger more inclusive clusters. These are methods which essentially produce a single partition as a final product. Major approaches within this category follow.

(1) *Total ennumeration of partitions and related methods.* Fortier and Solomon (1966) described an attempt at looking into total enumeration of all clustering partitions of a data set (developed by Kahl and Davis, 1955) and then selecting the best clustering by the use of an appropriate index. The number of distinct partitions of n elements into k groups is the so-called Stirling Number of the Second Kind. This number is so large, even for small n and k, that a total enumeration procedure appears to be unfeasible even for modest data sets. As an alternative, Fortier and Solomon tried random sampling of the total partitions for $n=19$, $k=2,3,\ldots,18$ (10 000 partitions obtained with a computer by unrestricted random sampling for each fixed size, k, of number of clusters). The results were disappointing by comparison with those obtained through Tryon's (1939) clustering

technique, using the 'coefficient of belonging' (Holzinger and Harman, 1941) as evaluation criterion. Although neither the total enumeration procedure nor the random sampling of the universe of partitions seem to be promising (Solomon, 1971), they represent logical steps in the historical development of quantitative taxonomy.

(2) *Nearest centroid sorting methods.* The basic procedure in the nearest centroid sorting methods is the selection of seed points to be used as cluster nuclei around which the set of entites can be grouped.

(i) *Methods with fixed number of clusters.* Here, k starting points are used as initial estimates of cluster centroids, corresponding to the specified number of clusters. Entities are then allocated to the cluster to whose centroid they are nearest. In the methods developed by Forgy (1965) and Jancey (1966) the estimates of the centroids are updated after each full cycle of allocation of entities. By contrast, in the 'k-means' method developed by MacQueen (1967) the cluster centroid is updated after the addition of each entity to the cluster.

(ii) *Methods with variable number of clusters.* These methods are elaborations on the basic nearest centroid sorting process, allowing the number of clusters to change (presumably to conform to the natural structure of the data) during the allocation process by forcing the division of a cluster with a large within-cluster sum of squares and the combination of two clusters with small between-cluster sum of squares. This is achieved by using parameters of 'coarsening' and 'refinement' set by the user.

(3) *Reallocation methods using variance-covariance criteria.* The basic procedure in this category is to reallocate entities among a set of clusters in such a way as to optimize some discriminant function or variance-covariance criterion, in effect, to select partitions that maximize intercluster distances and minimize intracluster distances. Most of these criteria are based on the matrix equation

$$\mathbf{T} = \mathbf{W} + \mathbf{B}$$

where **T** is the total scatter or dispersion matrix, **W** is the matrix of within-clusters dispersion, and **B** is the between-clusters dispersion matrix. We seek the situation where **W** is small and equivalently **B** is large in some sense, e.g. their determinantal values are small and large, respectively.

(4) *Density search methods.* These methods look for clusters in a set of points depicted in a metric space by seeking regions of high point-density separated by regions of low density. The most important of the methods in

this category is pattern clustering by multivariate mixture analysis, developed by Wolfe (1965, 1967, 1970). This method formulates the clustering problem as the decomposition of a presumed mixture of multivariate distributions using a maximum likelihood method. Wolfe's approach is quite elegant and appealing because of its underlying distribution theory. However, this is both its strength and its weakness, as in some situations the multivariate normal assumption may be quite inappropriate. A different type of method in this section is the taxmap method (Carmichael, George and Julius, 1968; Carmichael and Sneath, 1969) which is addressed to the detection of clusters in two or three dimensions by comparing relative distances between points and searching for continuous relatively dense regions of the space surrounded by continuous relatively empty regions. Another density searching method is that of Cattell and Coulter (1966), based on partitioning the multidimensional space and counting the number of points in each hypercube. Wishart's (1969a, b) mode analysis searches for natural clusters by estimating disjoint density surfaces in the sample distribution.

(5) *Other non-hierarchical methods.* The first approach to be mentioned in this miscellaneous section is Jardine and Sibson's (1968) overlapping clustering method, which consists in representing each point by a node on a graph and connecting all pairs of nodes which correspond to entities having a similarity of at least some preset value. Then a search is undertaken for the largest subsets of entites for which all pairs of nodes are connected. A different approach is that of Ling (1972, 1973) proposing a probabilistic theory of cluster analysis for finding (k,r)-clusters. These are generalizations of single linkage clusters such that each (k,r)-cluster has each of its elements within a distance r of at least k other elements of the same cluster and the entire set can be linked by a chain of links, each less than or equal to r. A third approach in this section is simultaneous clustering of variables and entities. For example, Lambert and Williams (1962) cluster entities on the basis of intervariable associations, and variables on the basis of interentity associations, allowing each process to be modified by the other. Similar procedures have also been proposed by Good (1965) and Hartigan (1972, 1975). Degerman (1970) described a two-stage method. First, clusters are found using an optimization technique whose clustering criterion is based on intracluster and intercluster indices. Second, two types of orthogonal variation are determined, one due to class or group structure and the other due to factors orthogonal to the class structure.

As discussed above, a large variety of clustering procedures has been developed during the last several years, reflecting an increasing interest in many fields, most recently in the behavioral sciences, for quantitatively exploring the presence of clusters and other structures in data. However, little research evaluating these methods has been carried out. The few evaluative studies reported in the literature have compared few representative quantitative taxonomic methods, applied to one type of generic data and according to one or two evaluative criteria. We are therefore in a situation where there is a wild proliferation of grouping methods and yet only scarce information about rational ways for selecting and using them. The present empirical study was designed in an attempt to respond to this challenging situation. Its basic objective was to evaluate comparatively, according to several criteria, the performance of major representative quantitative taxonomic methods, using data sets from various fields of application. The incipient knowledge available about the comparative merits of relationship measures and clustering techniques and about their field specificity made it necessary to apply all of them systematically across the board. An additional objective was the investigation of the interrelationships of the quantitative taxonomic methods and their hierarchical organization.

Accomplishing these objectives involved, first, the selection of various specific clustering techniques representing the major approaches included in the conceptual classification of quantitative taxonomic methods outlined above. As a corollary feature, various relationship measures were chosen to be used in association with the clustering techniques. Second, the design of the study involved the selection of several evaluative criteria, as well as the implementation of pertinent quantitative indices for criterion measurement. Finally, data sets from four different fields (psychosocial, psychopathological, anthropological and botanical) were obtained in order to run the taxonomic methods and to examine their specificity and generalizability. Artificial data sets were not used in this study as they are useful for a different although complementary purpose, namely, the clarification of the relationship between clustering procedure and concept of cluster implied by the procedure. Detailed descriptions of the data sets are presented in the following chapters, preceding the results and discussion of their taxonomic analyses.

2.2 QUANTITATIVE TAXONOMIC METHODS

In the present study, a quantitative taxonomic method is identified by the combination of a specific clustering procedure and a relationship measure (if the method requires this).

Three types of relationship measures were selected. These were the city-block distance (in which the distance between two points is the sum of the absolute value of their differences on each variable), the Euclidean distance (in which the square of the distance from one point to another is the sum of their squared differences on each variable), and the product-moment correlation coefficient. They were selected because of their frequent taxonomic use and the variety of relationship aspects they, as a group, evaluate. Interprofile similarity is evaluated by the product-moment correlation coefficient exclusively regarding profile shape, and by the two distance functions regarding different combinations of profile level, dispersion and shape.

Ten clustering methods were selected as representatives of the major approaches, in such a way that the classification of quantitative taxonomic methods presented in the previous section was covered fairly well. In order to implement these methods, clustering computer programs were obtained from various parts of the United States and made operative at the IBM 360/67 computer system of the Stanford Center for Information Processing. This library of clustering algorithms was complemented with a master program able to prepare the input data in three different forms (raw data, standardized data and factor scores), to compute matrices from these data corresponding to the three different relationship measures mentioned above, and to coordinate the use of each of the clustering algorithms with each of the relationship matrices, as needed.

Table 2.1 lists the 18 quantitative taxonomic methods evaluated in the present study, each defined by a clustering technique and a relationship measure, if any. Clustering procedures are labeled by numbers (from 1 to 10) and relationship measures by letters ('A' for correlation coefficient, 'B' for Euclidean distance, 'C' for city-block distance and 'X' for none used). When possible all relationship measures are employed in the comparative analysis between and within taxonomic methods, even though in some situations it may seem out of place to employ correlation coefficients or distances.

A brief description of the quantitative taxonomic methods used in the present study follows. The methods are ordered according to their clustering procedure components, including in the description the relationship measure used in combination with the particular clustering technique.

2.2.1 *Wherry-Wherry Q-Factor Analysis*

This algorithm represented the Q-factor analysis taxonomic method, which is one of the data representation (unfinished clustering) methods using a relationship matrix. The only relationship measure systematically used in this approach was the product-moment correlation coefficient. The Wherry-Wherry hierarchical factor analysis program (Wherry and Wherry,

Table 2.1
Quantitative taxonomic methods evaluated in the present study

Symbol	Relationship measure	Clustering algorithm	Taxonomic approach	Reference
1.A	Correlation coefficient	Wh-Wh factor analysis	Q-factor analysis	Wherry and Wherry (1971)
2.A	Correlation coefficient	MDSCAL-5M	Ordinal multidimensional scaling	Kruskal (1964a, b)
2.B	Euclidean distance	MDSCAL-5M	Ordinal multidimensional scaling	Kruskal (1964a, b)
2.C	City-block distance	MDSCAL-5M	Ordinal multidimensional scaling	Kruskal (1964a, b)
3.X	—	Chernoff's faces	Data representation without relationship matrix	Chernoff (1973)
4.A	Correlation coefficient	Single linkage HICLUS	Hierarchical agglomerative single linkage	Johnson (1967)
4.B	Euclidean distance	Single linkage HICLUS	Hierarchical agglomerative single linkage	Johnson (1967)
4.C	City-block distance	Single linkage HICLUS	Hierarchical agglomerative single linkage	Johnson (1967)
5.A	Correlation coefficient	Complete linkage HICLUS	Hierarchical agglomerative complete linkage	Johnson (1967)
5.B	Euclidean distance	Complete linkage HICLUS	Hierarchical agglomerative complete linkage	Johnson (1967)
5.C	City-block distance	Complete linkage HICLUS	Hierarchical agglomerative complete linkage	Johnson (1967)
6.A	Correlation coefficient	King's centroid average linkage	Hierarchical agglomerative average (centroid) linkage	King (1967)
7.A	Correlation coefficient	*k*-Means	Nearest centroid sorting with fixed number of clusters	MacQueen (1967)
7.B	Euclidean distance	*k*-Means	Nearest centroid sorting with fixed number of clusters	MacQueen (1967)
7.C	City-block distance	*k*-means	Nearest centroid sorting with fixed number of clusters	MacQueen (1967)
8.X	Several	ISODATA	Nearest centroid sorting with variable number of clusters	Ball and Hall (1965, 1967)
9.X	—	Rubin-Friedman covariance criterion optimization	Reallocation using variance-covariance criteria	Rubin and Friedman (1967)
10.X	—	NORMAP/NORMIX	Density search method	Wolfe (1971)

1971) employs both the principal factor and minimum residual methods to obtain basic factors which are then rotated to the Varimax solution. The Varimax factors can be further analyzed to yield a hierarchical factor structure which attempts to maximize simple structure. The hierarchical rotation was not used in this study. The cluster configuration was elucidated by inspection of the Varimax factor matrix. Each entity was assigned to the factor (cluster) on which it had the highest factor loading. The determination of cluster configuration by examining the factor matrix seems to be more effective than the plotting of points on a two-dimensional space (using two factors as axes) for visual clustering (Overall and Klett, 1972, p. 201).

2.2.2 *MDSCAL (version 5M)*

MDSCAL is a representative algorithm for performing ordinal multidimensional scaling which, from a clustering viewpoint, is a data representation (unfinished clustering) method using a relationship matrix. The guiding criterion (minimization of 'stress' value) in the MDSCAL process is to increase monotonicity between the ranking of interpoint distances in the original data and the corresponding ranking in the created multidimensional space. Details about version 1 of MDSCAL can be found in Kruskal (1964a, b). A recent version (5M) of this algorithm, available at the Stanford Center for Information Processing, was used in the present study. Three different quantitative taxonomic methods were obtained by alternatively using correlation coefficients, Euclidean distances and city-block distances with MDSCAL. The starting configuration used in every run was a two-dimensional random configuration supplied by the program. For determination of the cluster configuration, all entities were plotted in a two-dimensional space using two-coordinate solutions produced by MDSCAL. It has been shown (Shepard, Romney and Nerlov, 1972) that this technique is particularly powerful for facilitating visualization of patterns in a few dimensions. Two judges inspected the graphs and formed clusters subjectively. They were asked to use the following concept of a cluster (Everitt, 1974, p. 44): a continuous region of the space containing a relatively high density of points, separated from other such regions by regions containing a relatively low density of points (a definition which attempts to overcome the restrictions of the traditionally spherical concept of a cluster).

2.2.3 *Chernoff's Faces*

The algorithm produced by Chernoff (1973) to represent multivariate points through computer-drawn faces is, from a clustering viewpoint, the best known example of a data representation taxonomic method not using an explicit relationship matrix. The determination of cluster configuration in this case is based on groupings of faces by judges presumably using their abilities to evaluate configural information. Two layman judges, who had in a previous task performed better than four other judges in correctly classifying faces which represented aritificial entites designed to constitute three clearly different groups, were selected for participation in this study. The same two judges inspected and clustered the Chernoff faces and the points in the MDSCAL plots.

2.2.4 *Single Linkage HICLUS*

This algorithm, developed by Johnson (1967) with the name of 'minimum' or 'connected' technique, was used in the present study as representative of the single linkage agglomerative hierarchical clustering method. Three taxonomic methods resulted by alternatively using correlation coefficients, Euclidean distances and city-block distances with this clustering technique.

2.2.5 *Complete Linkage HICLUS*

This is the 'maximum' or 'diameter' technique developed by Johnson and presented in conjunction with the single linkage procedure in his well-known paper on hierarchical clustering schemes (1967). This algorithm was used in the present study as representative of the complete agglomerative hierarchical clustering method. Correlation coefficients, Euclidean distances and city-block distances were alternatively used with this clustering procedure.

2.2.6 *King's Centroid Average Linkage Technique*

This algorithm, described by King (1967) in his paper on clustering procedures, was used in the present study as representative of the centroid average linkage agglomerative hierarchical clustering method. It used the Pearson product-moment correlation coefficient as a relationship measure, but distance measures can also be employed as similarity indices in this procedure. We now go into it in some detail because of the prominence it receives later, and in order to review a clustering process concretely.

The procedure proposed by King is a step-wise clustering procedure. This is its principal asset because it leads to a simple and quick algorithm that involves $(n-1)$ scannings of a correlation matrix based on n entities. At each scanning or pass, the entities are sorted into a number of groups that is one less than at the previous pass. In this way, we obtain $(n-k)$ groups of entities at the k-th scanning. The $(n \times n)$ matrix can also be a distance matrix.

The procedure operates as follows. We will employ the correlation matrix as our similarity matrix for expository purposes, and bring in the distance matrix when appropriate to highlight differences.

As a start, we can view the *n* entities as *n* groups (one entity to each group) and then scan the correlation matrix for the maximum cell entry. In a distance matrix we would seek the minimum distance cell entry. Suppose the maximum correlation is between entities X_i and X_j. Label it r_{ij}. We place X_i and X_j in the same group, and we now have $(n—1)$ groups $X_1,X_2,\ldots,(X_i,X_j),\ldots,X_{n-1},X_n$. This produces an $(n—1) \times (n—1)$ correlation matrix, all pairs of correlation coefficients over the original $(n—2)$ entities plus the correlations obtained by pairing each of these with the concocted entity $X_i + X_j = Y_{ij}$. Essentially, we are representing the group of two entities by its centroid.

On the second pass of what is now an $(n—1) \times (n—1)$ correlation matrix, a third entity may join the group of two entities formed on the first pass if the correlation between it and Y_{ij} is maximum, or the maximum correlation value in the reduced correlation matrix may again involve two individual entities. Thus we would obtain either one group of three entities and $(n—3)$ groups each containing one entity, or two groups each containing two entities and $(n—4)$ groups each containing one entity. In either situation we merge entities and revise the correlation matrix as on the first pass. In the former case, the centroid of the group of three entities represents its group, and in the latter case, each group with two entities is represented by its centroid. Recall that we do not have to divide the sum of the entities by the number of entities to obtain the centroid because the correlation coefficient is invariant when one entity of the pair is always multiplied by the same constant.

Thus, at each pass the two groups with the highest correlations are merged and the total number of groups to that point is reduced by one, unless there are ties. After an entity has joined a group of variables, it cannot be removed from that group. In this way it is possible to miss an optimal grouping. This is very similar to the selection of predictors in stepwise linear regression. It should also be mentioned that a group can lose its identity by merging with another group on a later pass. By the time all the scanning is completed we have produced successively $(n—1),(n—2),(n—3),\ldots,3,2$ groupings.

The clustering index employed by King for measuring the worth of the grouping is that of minimum correlation (or maximal distance) between the group centroids when the scanning has placed the entities into two groups. This leaves something to be desired because it does not look at the effectiveness of the grouping when more than two groups are involved. He also reviews another index, suggested originally by Wilks, for testing the mutual independence of k subsets of n multivariate normal random entities. In terms of what we described earlier, the index is the ratio of the determinants determinants

$$Z = \frac{|\mathbf{T}|}{\prod_{i=1}^{k} |\mathbf{W}_i|}$$

where $\mathbf{T}$ is the scatter matrix defined previously and each $\mathbf{W}_i$ is the scatter matrix for each of the k groups.

This index has some nice geometrical and statistical properties. For example, when $k = 2$,

$$Z = \frac{|\mathbf{T}|}{|W_1| \cdot |W_2|} = \pi (1 - r_i^2)$$

where r_i is the i-th canonical correlation between the two sets of variables. This index may be viewed as a 'generalized alienation coefficient' since it is an extension of $1 - R^2$, where R is the multiple correlation coefficient occurring when two groups have one variable in one group and $(n - 1)$ in the other. However, it is not too useful in some data analyses, especially in social science, because a number of data sets lead to quasi-singular correlation matrices and truncation error can give ridiculous results. For this reason, and possibly others, negative determinants appear and make it impossible to employ the Wilks index.

2.2.7 *k-Means Convergent Technique*

This technique is a refinement of the basic k-means method described by MacQueen (1967), which is probably the fastest of all explicit clustering methods. Convergence refers here to the iteration of reallocation cycles until a full cycle completed through the data set fails to cause any changes in cluster membership. The algorithm implementing this technique was obtained from Anderberg (1973). It was used in the present study as representative of the non-hierarchical nearest centroid clustering approach with fixed number of clusters. This technique was alternatively used with correlation coefficients, Euclidean distances and city-block distances.

2.2.8 *ISODATA*

ISODATA, the most sophisticated representative of the non-hierarchical nearest centroid clustering approach with variable number of clusters, was developed by Ball and Hall (1965, 1967) at the Stanford Research Institute.

In the version used in the present study, the algorithm first computes an overall mean vector as the center of one very large cluster. All points whose distance from this centroid is greater than a preset 'sphere value' are assigned to clusters outside the large mean cluster. Then several *k*-means type of settling iterations take place. Next, a lumping and splitting iteration process, controlled by preset values, is carried out. After this, new cluster centroids are computed and the entities are accordingly reallocated. The entire process iterates ten times. Because in this study we required a partition with a certain number of clusters, the values controlling the clustering process were manipulated in such a way that the process ended with the desired number of clusters. ISODATA uses a squared Euclidean distance as an index of relationship for the *k*-means part of the procedure, as well as a Euclidean distance in the lumping process and the within-group variance of the variables in the splitting process.

2.2.9 *Rubin and Friedman's Covariance Criterion Optimization Technique*

This is the 'hill-climbing' algorithm described by Rubin and Friedman (1967) to sift through the partitions of a set of entities seeking to optimize one of several variance-covariance criteria (Friedman and Rubin, 1967). The criterion used in the present study was the minimization of Wilks' lambda, which corresponds to the ratio of determinants $|\mathbf{W}|/|\mathbf{T}|$. This criterion proved to be the most effective in cluster analyses carried out by Friedman and Rubin (1967), who recommended it for the exploration of the structure of heterogeneous multivariate data. Rubin and Friedman's covariance criterion optimization technique was used in the present study as representative of the non-hierarchical reallocation approach using variance-covariance criteria. It does not use any of our conventional relationship measures.

2.2.10 *NORMAP/NORMIX*

This approach, developed by Wolfe (1971) under the name of pattern clustering by multivariate mixture analysis, is based on the assumption that the population under consideration is composed of several classes each having a multivariate normal distribution. The problem is formulated as the decomposition of the mixture of distributions by estimating their parameters through a maximum likelihood method. Entities are then assigned to clusters for which their probability of belonging is greatest. NORMAP specifically deals with the case in which equal within-group covariance matrices are assumed; NORMIX deals with the non-restrictive case. Wolfe's method was used in the present study as representative of the density search non-hierarchical approach. It does not use any of our conventional relationship measures.

2.3 EVALUATIVE CRITERIA

Several criteria and indices for evaluation were used in the present study for assessing the performance of the 18 quantitative taxonomic methods applied to data sets from four different fields. The three major criteria, used for evaluating all taxonomic methods, were external criterion validity, internal criterion validity and replicability.

2.3.1 *External Criterion Validity*

The external criterion validity of a given quantitative taxonomic method was assessed by comparing its resulting cluster configuration with a cluster configuration or classification previously established by field experts on a given data set (in order to make the analysis commensurate, the cluster configuration produced by quantitative methods was required to have the same number of clusters as the expert's cluster configuration). The degree of similarity between these configurations was determined through the computation of several statistics, divided into two groups as follows.

(1) *Statistics computed for a contingency table displaying the cross-classification of entities in both cluster configurations.* The first statistic of this group was the percentage of concordance between the quantitative method and the expert. This is defined as 100 times the ratio of the number of agreements to the total number of entities. The second statistic was χ^2 which tested the hypothesis of independence between configurations. The p value associated with χ^2 (probability of error in rejecting a true null hypothesis) can also be used as an inverse index of association between configurations. As this index is adjusted for the number of columns and rows, it can be useful for comparing results from contingency tables of different sizes. However, its use in comparative tasks becomes cumbersome and inaccurate when it attains very small values. Other χ^2-related statistics of overall association between configurations used in the present study were Cramér's (1946) statistic and the contingency coefficient. It should be noted that p values, contingency coefficients and Cramér's statistic are monotone with each other and monotonicity is all that is needed for ranking. Although all these three indices were computed, only the Cramér's statistic results were used for ranking the taxonomic techniques.

(2) *A coefficient of correlation between corresponding entries of two matrices of similarities between entities.* These matrices were derived, respectively, from the cluster configuration produced by the quantitative method under examination and the cluster configuration established by the expert. The entries of each similarity matrix were determined to be 1 if the corresponding entities belonged to the same cluster, and 0 otherwise.

As there is no universally accepted index of external validity in this context, it seemed to be worthwhile to compute and compare the various statistics mentioned above.

2.3.2 *Internal Criterion Validity*

The internal criterion validity of a given quantitative taxonomic method was conceptualized as the appropriateness of its resulting cluster configuration to the original structure of the data. This validity was estimated through the cophenetic correlation coefficient introduced by Sokal and Rohlf (1962). This is a product-moment correlation coefficient computed between corresponding entries of a similarity matrix derived from the cluster configuration and the initial similarity matrix. The entries of the derived similarity matrix were determined to be 1 if the entities involved belonged to the same cluster, and 0 otherwise.

2.3.3 *Replicability and Stability*

In order to assess the replicability of the cluster configurations produced by a given quantitative taxonomic method, two data sets were developed by randomly and equally dividing, within each expert-produced cluster, the total group of entities available from each data base. The taxonomic method under examination was applied to these two 'equivalent' data sets, and the degree of similarity between the two resulting cluster configurations was computed as an index of replicability. In our study, the only labeling of entities valid across the 'equivalent' data sets was determined by the cluster to which an entity was assigned by the expert. Therefore, the assessment of similarity between the two resulting cluster configurations, was made through the computation of a correlation coefficient between corresponding entries of two contingency tables, each depicting the cross-classification of the cluster configurations produced by the quantitative method and the expert on one of the 'equivalent' data sets.

In the case of the ethnic populations data, the replicability of cluster configurations produced by a given quantitative taxonomic method was studied in terms of their stability across two different sets of variables. The data sets in this case were composed of the same ten entities, but measured on two different sets of 29 variables developed by randomly halving the original set of 58 variables. An estimate of the degree of similarity (stability index) between the two resulting cluster configurations was obtained by computing a correlation coefficient between corresponding entries of the two expert-method cross-classification tables from data sets A and B.

2.3.4 *Inter-rater Reliability*

As the clusterings obtained through the three MDSCAL methods and the Chernoff's faces method involved the participation of human judges, it was pertinent to assess their inter-rater reliability. This was done by comparing the corresponding configurations completed by the two judges upon their examining the graphical displays. The statistics used were those described in the external criterion validity section for assessing agreement between expert and quantitative method.

2.4 CLUSTER ANALYSIS OF QUANTITATIVE TAXONOMIC METHODS

A cluster analytic study of the quantitative taxonomic methods was undertaken on the basis of the similarities of their performance on each of several data sets.

The procedure followed involved, first, the derivation of an interentity similarity matrix from each of the cluster configurations produced by all quantitative taxonomic methods on each data set. The entries of this matrix were determined to be 1 if the entities involved belonged to the same cluster, and 0 otherwise. Then, proximity measures between pairs of taxonomic methods were obtained by computing a coefficient of correlation between corresponding entries of their derived similarity matrices on that data set.

A mean of the four correlation coefficients for the same pair of taxonomic methods was computed across the first half of each of the four original data bases, and the same was done for the second halves. In this way, two matrices of mean intercorrelations among taxonomic methods were obtained across data bases. Each matrix was cluster analyzed using taxonomic procedures selected on the basis of their overall performance up to that point in the study.

3

Cluster Analysis of Treatment Environments

3.1 INTRODUCTION

One of the most interesting developments of the last two decades in psychiatry and behavioral sciences has been the systematic study of the social climate of treatment settings. Treatment environments seem to have significant influence on patients' behavior (Cumming and Cumming, 1957). It has also been well documented that psychiatric treatment programs vary considerably in terms of social atmosphere characteristics (Moos, 1974a, b) and that these differences appear to be related to treatment outcome (Moos, Shelton and Petty, 1973).

To increase explanatory and predictive powers in behavioral sciences, a 'treatment by abilities' design has been cogently proposed by Cronbach (1975). This idea has been implemented in clinical psychology and psychiatry through attempts to match types of patient with types of treatment program in such a way as to optimize outcome. Such endeavor requires appropriate classifications of both patients (usually according to diagnostic categories) and treatment programs (for example, according to their milieu characteristics). A classification of inpatient treatment environments was produced by Price and Moos (1975) on both empirical and judgmental bases. They found that the groups obtained differed systematically in institutuional affiliation, program size and patient/staff ratios.

The first data base used in the present cluster analytic study consisted of a set of psychiatric inpatient wards characterized by their social climate features. Clusters of psychiatric wards were developed *de novo* by applying the quantitative taxonomic methods descibed in § 2.2. Some of the clustering processes and outcomes obtained are presented for illustrative purposes. Then, the comparative performances of the quantitative taxonomic methods and the individual relationship measures on the treatment environment data base are presented. The evaluative criteria used are described in § 2.3. Finally, a compendium of the levels and areas of agreement and disagreement between the clustering configuration provided by an expert in the field and those produced by the various quantitative taxonomic methods, is presented.

3.2 DATA BASE

Data on 72 psychiatric inpatient treatment environments measured on the ten subscales of the Ward Atmosphere Scale (Moos, 1974a) was kindly provided by Professor Rudolf Moos, Director of the Social Ecology Laboratory, Stanford University.

The Ward Atmosphere Scale is a self-reporting instrument that measures the consensual perception of the social climate of such settings by their inhabitants: patients and staff. Each of its ten subscales has ten items and is scored from 0 to 10 according to the number of items endorsed in the direction appropriate to the meaning of the scale. A brief description of the ten subscales follows:

(1) *Involvement.* Degree of patient activity in the daily functioning of the program.
(2) *Support.* Degree to which patients are encouraged and supported by staff and other patients.
(3) *Spontaneity.* Degree to which the program encourages patients to express themselves openly.
(4) *Autonomy.* Degree of self-sufficient decision-making encouraged in patients.
(5) *Practical orientation.* Degree to which the environment encourages patients to prepare for discharge.
(6) *Personal problem orientation.* Degree to which patients are encouraged to be concerned with and have an insight into personal problems.
(7) *Anger and aggression.* Degree to which patients are allowed and encouraged to argue and express aggressive behavior.
(8) *Order and organization.* Degree of importance of activity planning and neatness in the program.
(9) *Program clarity.* Clarity of goal expectations and rules.
(10) *Staff control.* Degree to which staff determines rules and their strictness.

The 72 treatment environments used in the present study were drawn from a larger pool collected by Moos (1974b), by his selecting 18 typical members from each of four different groups or types of treatment environments. These four types, labeled 'therapeutic community', 'relationship oriented', 'insight oriented' and 'control oriented', are components of a six-type classification of inpatient treatment environments developed by Price and Moos (1975). This classification was produced with the assistance of a single linkage hierarchical clustering procedure (Carlson, 1972) using a pool of 144 psychiatric wards obtained from state hospitals, Veterans Administration hospitals, university and teaching hospitals and community and private hospitals.

The 'therapeutic community' program type resembles Maxwell Jones' (1953) therapeutic community concept in that patient involvement and therapeutic orientation are emphasized. This program type does not seem to occur in Veterans Administration hospitals but is rather evenly distributed among other types of institutions. It seems to have a relatively small number of patients and low patient/staff ratios.

The 'relationship oriented' program type seems to occur in all types of institutions and is more frequently represented in Veterans Administration and university hospitals. As for 'therapeutic community' programs, this type seems to have a relatively small number of patients and low patient/staff ratios.

The 'insight oriented' program type seems to predominate in state hospitals according to Price and Moos' (1975) data, although it also appears in the other three types of institutions. Programs of this type have a relatively large number of patients and high patient/staff ratios.

The 'control oriented' program type is mainly found in Veterans Administration and state hospitals. As the 'insight oriented' program type, this type has a relatively large number of patients and high patient/staff ratios.

It can be noted that regarding institutional affiliation, program size and patient/staff ratio, the four types of treatment environment represented in the present study appear to form two pairs: one constituted by the 'therapeutic community' and 'relationship oriented' types, and the other by the 'insight oriented' and 'control oriented' types.

In order to examine the replicability of cluster configurations produced by the various quantitative taxonomic methods under evaluation, two similarly composed data sets were required. These two data sets, labeled A and B, were obtained from the total group of 72 treatment environments by randomly halving each of the four 18-member subgroups mentioned above.

Tables 3.1 and 3.2 present data matrices constituted by the averages of the patient and staff ratings on each of the ten Ward Atmosphere Subscales for each of the 36 treatment environments from data sets A and B, respectively. Table 3.3. presents, separately for data sets A and B, the

Table 3.1
Data matrix of treatment environments measured on ten Ward Atmosphere Subscales: data set A

Treatment environments		Ward Atmosphere Subscale 1	2	3	4	5	6	7	8	9	10
Therapeutic community	1	6·277	5·340	3·106	4·596	4·894	3·043	3·362	6·532	5·617	6·553
	2	5·804	5·543	3·500	4·065	6·109	4·522	5·217	6·174	5·326	6·826
	3	5·200	5·543	3·657	3·143	5·971	3·971	4·943	6·171	5·200	6·571
	4	6·111	5·194	3·139	5·806	5·917	3·806	3·806	6·611	5·944	7·083
	5	7·045	5·716	3·507	6·045	6·254	3·866	3·716	6·254	5·567	6·582
	6	6·871	5·774	3·419	4·839	6·129	3·290	2·968	8·258	5·710	6·613
	7	6·611	5·778	3·500	4·500	5·556	4·500	3·889	7·556	5·056	7·389
	8	6·571	6·429	3·286	4·429	5·857	4·571	4·286	6·714	5·571	6·571
	9	6·700	5·150	4·200	4·700	5·500	4·500	5·150	6·900	4·300	7·250
Relationship oriented	10	4·903	3·968	4·226	5·032	5·097	5·774	6·323	4·871	5·065	5·548
	11	6·603	4·444	4·270	4·746	6·095	6·429	6·032	4·603	5·048	6·825
	12	4·231	4·231	4·846	5·000	5·923	4·769	6·154	3·000	4·000	4·923
	13	4·500	6·400	4·100	4·700	5·600	5·700	7·100	4·300	4·700	6·400
	14	5·973	4·459	3·946	4·378	5·216	6·486	6·432	5·757	4·351	6·135
	15	4·556	5·056	3·333	4·722	6·222	3·833	4·778	5·389	4·778	7·000
	16	5·644	4·667	3·511	5·333	5·778	4·022	5·333	5·178	5·044	6·222
	17	4·857	4·857	5·762	4·143	5·286	5·333	6·143	4·048	4·952	4·476
	18	6·125	5·125	3·625	3·625	5·500	5·625	7·875	3·625	3·875	4·625

Insight oriented	19	5·375	6·250	4·750	4·500	5·625	4·250	4·625	6·250	6·500	3·750
	20	7·194	6·389	4·139	6·111	6·528	4·167	3·250	7·722	6·000	5·917
	21	8·000	6·583	4·875	5·208	6·417	4·708	3·667	7·875	6·125	4..458
	22	8·087	7·348	6·174	5·043	6·348	4·696	3·435	8·304	5·826	4·391
	23	8·192	7·615	5·654	5·769	8·000	5·115	1·962	9·192	6·885	5·615
	24	9·839	7·774	6·484	7·419	7·065	7·806	7·387	8·387	7·226	4·290
	25	5·714	7·286	7·143	6·286	5·429	4·714	4·143	6·429	8·286	3·429
	26	7·125	7·625	4·500	5·750	6·250	3·125	4·000	7·000	6·000	4·375
	27	6·000	5·467	6·600	4·800	5·600	5·400	5·467	3·600	6·200	2·667
Control oriented	28	4·800	5·700	6·900	4·400	6·200	6·800	7·000	3·800	5·100	3·500
	29	8·471	7·647	5·588	6·118	6·588	6·412	5·588	6·647	5·588	3·471
	30	9·021	6·146	6·208	5·938	6·833	7·313	7·354	6·021	5·958	4·208
	31	8·000	7·500	6·500	6·900	6·000	6·800	6·800	3·900	6·100	2·500
	32	7·273	6·000	4·909	5·182	7·273	4·727	4·364	6·273	5·091	5·364
	33	9·625	6·563	7·063	7·250	7·438	7·313	8·000	4·375	4·313	3·125
	34	6·471	5·765	6·529	4·706	5·647	6·529	5·353	3·765	3·824	3·235
	35	7·304	7·522	7·435	5·043	5·043	7·435	5·783	6·435	5·174	2·043
	36	8·404	5·519	5·327	5·308	6·462	6·615	6·269	6·019	4·885	4·538

Table 3.2
Data matrix of treatment environments measured on ten Ward Atmosphere Subscales: data set B

Treatment environments		Ward Atmosphere Subscale 1	2	3	4	5	6	7	8	9	10
Therapeutic community	1	7·179	5·893	4·250	5·893	7·000	5·357	3·429	7·321	6·393	7·143
	2	6·889	5·306	3·806	4·194	5·667	3·889	3·639	7·306	5·417	6·306
	3	5·813	4·667	3·042	5·688	5·479	3·979	4·563	5·625	4·604	6·792
	4	6·000	5·606	2·848	4·545	5·515	3·636	3·394	6·939	5·333	6·121
	5	6·731	5·962	3·385	4·808	5·423	3·538	3·115	8·308	5·462	6·923
	6	5·935	4·457	3·065	5·696	5·696	4·109	4·935	5·391	4·804	6·522
	7	7·708	5·944	3·306	5·875	6·278	4·611	3·875	7·375	6·097	6·708
	8	5·886	6·029	3·343	5·057	6·543	3·514	3·171	7·886	6·943	6·343
	9	6·750	4·750	3·500	5·125	5·688	3·438	4·875	6·125	4·938	7·438
Relationship oriented	10	5·032	4·097	3·742	4·548	4·419	5·452	6·161	4·774	4·774	6·032
	11	4·233	5·100	3·933	3·700	5·533	5·967	5·900	5·000	5·033	6·900
	12	6·433	4·299	4·299	4·791	6·448	6·567	6·582	4·851	5·179	6·597
	13	5·815	4·815	4·815	5·333	5·926	6·407	5·556	5·407	5·519	6·037
	14	5·455	4·818	3·545	4·455	5·182	5·000	6·000	4·909	4·273	6·364
	15	6·617	4·483	4·083	4·517	6·133	5·633	6·067	5·350	5·267	6·300
	16	7·708	4·369	5·062	4·538	6·000	6·815	6·508	5·785	4·600	6·215
	17	5·333	4·667	5·067	4·600	5·400	5·000	7·600	4·333	4·533	5·600
	18	4·688	4·917	4·417	5·250	5·583	3·583	4·750	5·250	5·208	6·000

Insight oriented	19	5·583	5·667	4·833	6·583	6·333	6·167	6·583	4·333	5·500	3·500
	20	5·667	5·667	6·000	7·000	5·500	4·667	4·500	4·333	5·167	3·500
	21	6·778	6·222	5·556	5·000	5·556	4·111	4·444	6·778	7·667	4·222
	22	5·529	5·765	5·235	4·529	5·706	5·059	3·941	5·588	6·059	3·471
	23	11·357	7·929	7·857	8·071	8·071	8·929	8·143	7·929	6·571	2·929
	24	6·750	7·125	4·125	5·625	7·250	4·250	5·000	6·500	6·125	3·375
	25	6·737	6·421	3·842	5·526	5·947	3·474	3·526	7·368	5·737	5·579
	26	6·100	5·600	4·900	6·400	6·350	6·150	6·450	4·500	5·400	3·500
	27	9·000	7·500	4·800	6·000	7·100	4·100	3·100	8·500	5·600	4·900
Control oriented	28	6·111	5·667	4·056	6·389	7·333	5·389	6·167	4·167	4·833	4·722
	29	8·067	6·867	5·733	6·933	7·400	5·867	7·267	6·467	6·200	3·333
	30	8·167	6·167	6·000	3·833	6·167	6·500	8·167	4·333	4·000	3·667
	31	7·952	6·429	5·476	6·429	7·286	6·952	5·762	8·048	5·333	3·857
	32	7·588	6·000	5·647	6·706	6·353	5·412	6·647	6·000	4·824	4·647
	33	8·308	6·077	6·000	5·462	7·615	4·769	6·538	5·692	6·692	4·077
	34	7·575	4·600	5·075	5·100	6·800	6·400	6·550	5·075	4·450	4·250
	35	7·833	7·000	4·333	6·333	7·333	4·833	5·167	6·833	4·833	5·167
	36	9·000	4·909	5·818	5·000	6·727	3·909	5·455	5·636	4·727	5·273

Table 3.3
Means and standard deviations of four groups of treatment environments, on the ten Ward Atmosphere Subscales, for data sets A and B

Treatment environment groups	Statistic	Ward Atmosphere Subscales 1	2	3	4	5	6	7	8	9	10
Data set A											
Therapeutic community	Mean	6·354	5·607	3·479	4·680	5·799	4·008	4·149	6·797	5·366	6·826
($N=9$)	SD	0·578	0·388	0·326	0·866	0·422	0·567	0·805	0·699	0·482	0·330
Relationship oriented	Mean	5·266	4·801	4·180	4·631	5·635	5·330	6·241	4·530	4·646	5·795
($N=9$)	SD	0·839	0·708	0·749	0·516	0·398	0·951	0·900	0·882	0·462	0·943
Insight oriented	Mean	7·281	6·926	5·591	5·654	6·362	4·887	4·215	7·195	6·561	4·321
($N=9$)	SD	1·472	0·789	1·066	0·893	0·807	1·273	1·532	1·652	0·792	1·007
Control oriented	Mean	7·708	6·485	6·273	5·649	6·387	6·660	6·279	5·248	5·115	3·554
($N=9$)	SD	1·453	0·857	0·845	0·974	0·764	0·815	1·127	1·249	0·731	1·022
Data set B											
Therapeutic community	Mean	6·543	5·402	3·394	5·209	5·921	4·008	3·888	6·920	5·555	6·700
($N=9$)	SD	0·670	0·628	0·426	0·616	0·554	0·626	0·721	1·001	0·779	0·426
Relationship oriented	Mean	5·702	4·618	4·329	4·637	5·625	5·603	6·125	5·073	4·932	6·227
($N=9$)	SD	1·075	0·327	0·559	0·478	0·600	1·001	0·776	0·425	0·408	0·376
Insight oriented	Mean	7·056	6·433	5·239	6·082	6·424	5·212	5·076	6·203	5·981	3·886
($N=9$)	SD	1·933	0·881	1·183	1·076	0·881	1·670	1·654	1·593	0·763	0·852
Control oriented	Mean	7·845	5·968	5·349	5·798	7·002	5·559	6·413	5·806	5·099	4·333
($N=9$)	SD	0·779	0·808	0·716	1·015	0·509	0·972	0·925	1·224	0·851	0·668

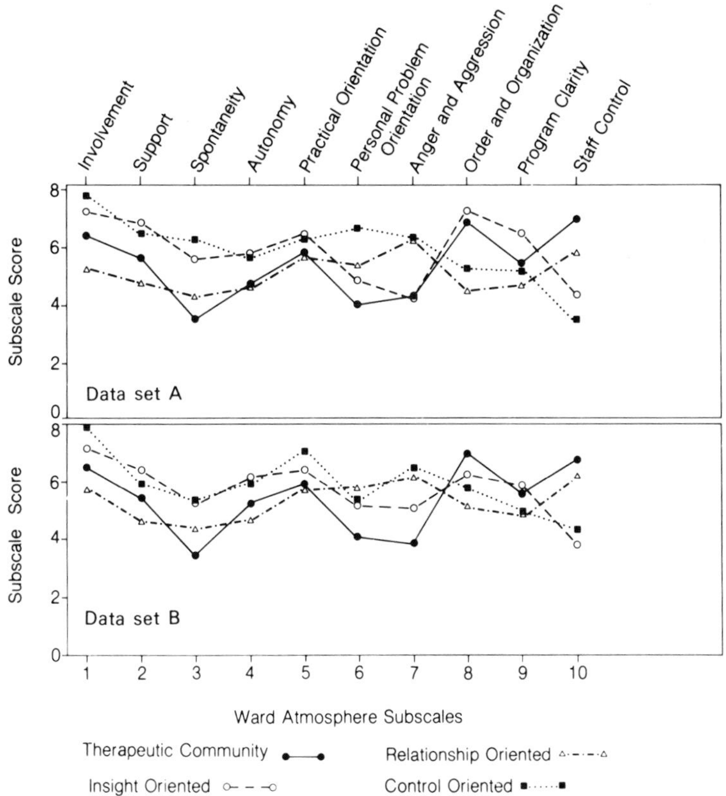

Fig. 3.1 Profiles of four treatment environment groups of data sets A and B.

means and standard deviations of the four groups of treatment environments on the ten Ward Atmosphere Subscales. Figure 3.1 exhibits the mean graphical profiles of the four treatment environment groups on data sets A and B.

3.3 ILLUSTRATIVE CLUSTER ANALYTIC RESULTS

The various quantitative taxonomic methods evaluated in this study and listed in Table 2.1 were applied to both data sets of treatment environments. In this section, a set of figures and tables will be presented to exhibit the clustering processes and results obtained from the application of some of these taxonomic methods. The intention is to provide examples of specific clustering outcomes obtained on this data base while illustrating the application of some of the taxonomic methods, which are not necessarily the best methods for treatment environments data. The ranking of the quantitative taxonomic methods on this data base will be presented in the next section.

Results from the application of the same two taxonomic methods will be presented for both data sets of treatment environments and for the other three data bases, in order to allow comparisons across fields. One of these two taxonomic methods is the complete linkage hierarchical method using correlation coefficients and, as shown in Chapter 7, this turned out to be the best ranked taxonomic method across evaluative criteria and fields of application. The other method is ordinal multidimensional scaling, also using correlation coefficients, which offers an interesting possibility for visualizing data structure. Additionally, two different taxonomic methods will be illustrated for each of the four data bases in order to cover the remaining eight basic clustering methods. The matching of taxonomic

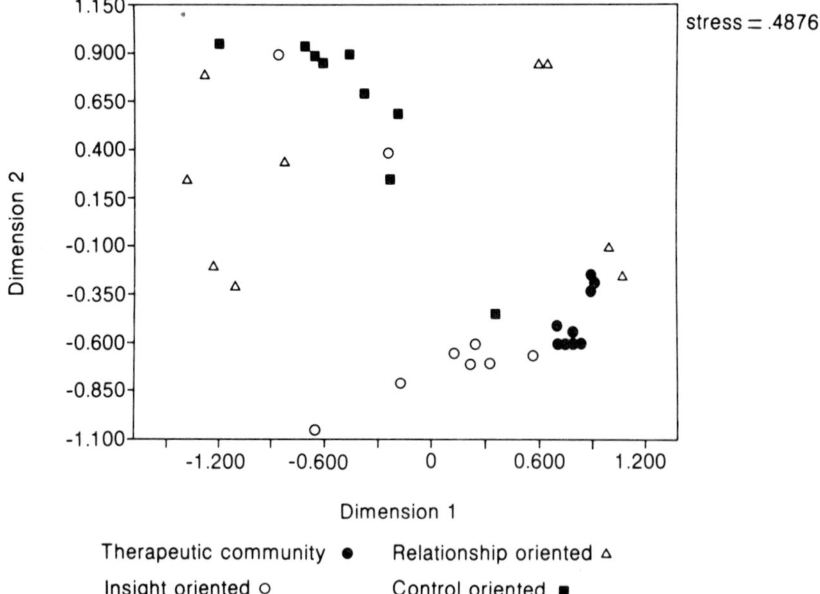

Fig. 3.2 Ordinal multidimensional scaling representation, using correlation coefficients, of the 36 treatments environments of data set A.

methods and data bases will be done in such a way as to present each method on the data base on which it had its best or second best ranking. Again, the particular methods presented for a given data base may not be the best methods for that or any other data base.

Figures 3.2 and 3.3 show two-dimensional representations of the treatment environments of data sets A and B, obtained by using ordinal multidimensional scaling and correlation coefficients. Each treatment environment is represented by a symbol indicating the group to which it belongs according to the expert. Disregarding the symbols, there does not seem to be a very neat cluster structure on these plots. The entities on data set A seem to some extent to be grouped into two clusters—one located on the upper left triangle and the other on the lower right triangle of the field—but such a dichotomous structure does not appear on data set B. By paying attention to the symbols, it is possible to see that most of the entities corresponding to the same groups according to the expert, clump together. This is particularly true for the 'therapeutic community' group, all of whose members appear packed together on both sets.

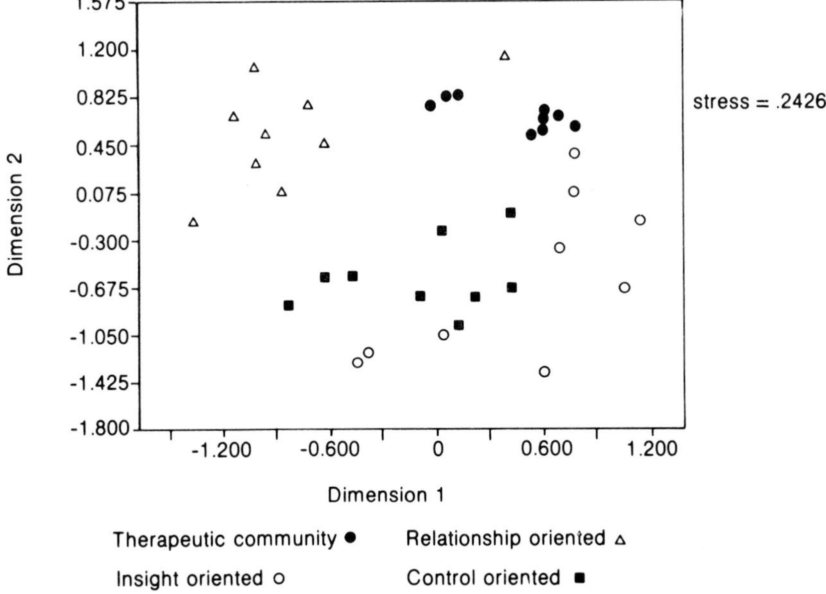

Fig. 3.3 Ordinal multidimensional scaling representation, using correlation coefficients, of the 36 treatment environments of data set B.

Figures 3.4 and 3.5 exhibit dendrograms representing complete linkage agglomerative cluster analyses using correlation coefficients, of the treatment environments of data sets A and B. For each data set, the whole hierarchical clustering process is displayed, from the beginning at the top, where each treatment environment is a single cluster, to the end where all treatment environments are lumped into one cluster. The four-cluster solutions, obtained at similarity levels —0·032 in data set A and —0·026 in data set B, showed noticeable agreement with the groups established by the expert. This was more clearly the case for 'therapeutic community' and 'relationship oriented'. The other two groups, 'insight oriented' and 'control oriented', although recognizable as the last two groups in both dendrograms, tended to mix with each other. In the next agglomerative step, these two groups were lumped into one cluster in both data sets.

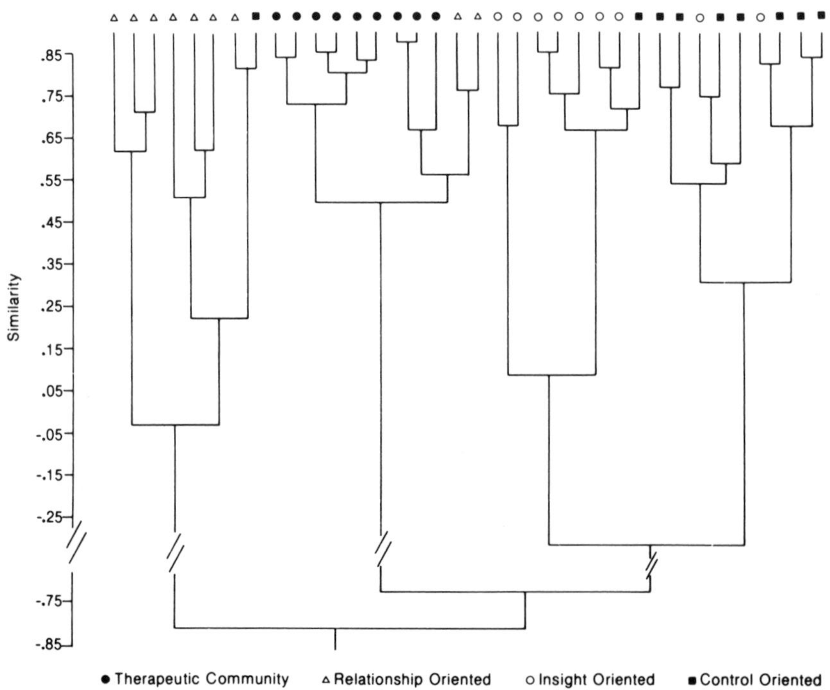

Fig. 3.4 Dendrogram representing a complete linkage cluster analysis, using correlation coefficients, of the 36 treatment environments of data set A.

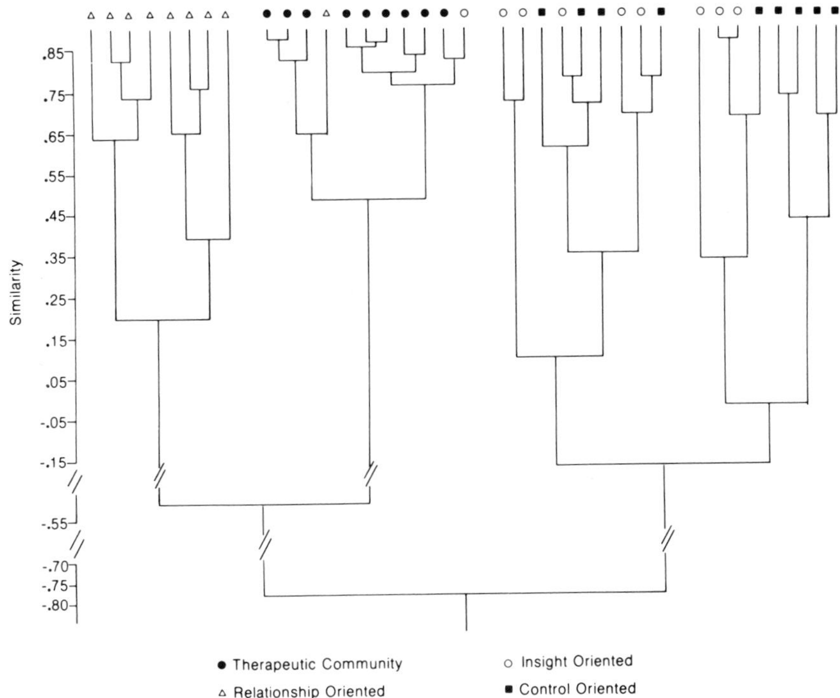

Fig. 3.5 Dendrogram representing a complete linkage cluster analysis, using correlation coefficients, of the 36 treatment environments of data set B.

Table 3.4 presents the basic steps of the convergent *k*-means cluster analysis, using correlation coefficients, of the 36 treatment environments of data set A. This is a nearest centroid sorting method with a fixed number of clusters, which was set to be four in order to correspond to the number of clusters established by the field expert. First, four treatment environments were randomly selected to serve as initial cluster seed points. Each of the remainder treatment environments were allocated to the cluster with the seed point closest to it. After this allocation cycle, new cluster centroids were computed by averaging the vectors of all members of each cluster. Then, in the first iteration, each of the treatment environments were reallocated, as necessary, to the cluster with the centroid closest to it. This resulted in five treatment environments being moved to different clusters,

Table 3.4
k-Means cluster analysis, using correlation coefficients, of the 36 treatment environments of data set A

Initial cluster seed points produced by a random number generator:
5 15 7 11

Iteration 1		
5 entities moved	Summed deviations about seed points:	16·81
Iteration 2		
4 entities moved	Summed deviations about seed points:	15·19
Iteration 3		
2 entities moved	Summed deviations about seed points:	13·84
Iteration 4		
0 entities moved	Summed deviations about seed points:	13·77

Final cluster membership

Cluster 1 contains 8 data units:	19 21 22 23 24 25 26 29
Cluster 2 contains 8 data units:	2 3 10 11 13 14 15 16
Cluster 3 contains 9 data units:	1 4 5 6 7 8 9 20 32
Cluster 4 contains 11 data units:	12 17 18 27 28 30 31 33 34 35 36

and the deviations about seed points summing to 16·81. New iterations were conducted, which progressively resulted in lower summed deviations about new seed points and in fewer entities being moved from one cluster to another until, in interation four, no entities changed clusters. Each of the final clusters included six or seven of the nine treatment environments corresponding to each of the groups established by the field expert.

Table 3.5 presents the Q-factor matrix for treatment environments of data set B, obtained through Q-factor analysis using the principal axis method and a Varimax rotation. Four factors were extracted because this was the number of groups of treatment environments established by the field expert for this data base. In order to determine cluster membership, each treatment environment was allocated to the Q-factor on which it had the highest factor loading, as marked. Although a neat cluster structure was not obtained, there appeared to be significant overlapping between each Q-factor and one of the expert-based groups. This was particularly so in the case of Q-factors 1 and 3 which correspond to the 'therapeutic community' and 'relationship oriented' groups, respectively.

Some general points about the cluster structure of the treatment environments data base can be formulated from the various taxonomic results presented above. This data base does not appear to have a very neat cluster

Table 3.5
Q-factor matrix for treatment environments of data set B

Treatment environments		Q-factors 1	2	3	4
Therapeutic community	1	0·5755	—0·1800	0·1678	0·6154
	2	0·4617	—0·2052	0·2467	0·7676
	3	0·7565	—0·0894	0·5085	0·1686
	4	0·6429	—0·1953	0·1907	0·6228
	5	0·5747	—0·2886	0·1307	0·6685
	6	0·7271	0·0336	0·5758	0·1238
	7	0·6264	—0·0314	0·2374	0·6373
	8	0·6535	—0·2370	—0·0188	0·5840
	9	0·6419	—0·1401	0·5288	0·3321
Relationship oriented	10	—0·0468	—0·1540	0·8137	—0·2259
	11	—0·0080	—0·4051	0·6498	—0·1955
	12	—0·1035	0·1455	0·8899	—0·0768
	13	—0·0688	—0·0188	0·7183	0·0179
	14	0·2310	—0·0270	0·8709	—0·0810
	15	0·0753	0·1276	0·8881	0·2251
	16	—0·3315	0·2068	0·7952	0·2919
	17	—0·2246	0·2488	0·5677	—0·4328
	18	0·8023	—0·2451	0·1332	0·0020
Insight oriented	19	—0·1130	0·7934	—0·1021	—0·3158
	20	—0·1098	—0·5677	0·6326	0·0788
	21	0·1930	0·0665	—0·4136	0·6473
	22	—0·0613	0·2385	—0·5152	0·6015
	23	—0·3444	0·7594	—0·1216	0·3853
	24	0·3758	0·5393	—0·2387	0·5077
	25	0·6586	—0·0114	—0·0827	0·6731
	26	—0·1415	0·8662	—0·0537	—0·1778
	27	0·4215	0·2136	—0·1511	0·7920
Control oriented	28	0·3003	0·7623	0·2559	—0·1971
	29	—0·0102	0·8722	—0·1536	0·2637
	30	—0·4999	0·6199	0·2912	0·0881
	31	—0·0479	0·5623	—0·0529	0·5854
	32	0·0664	0·8149	0·0793	0·2014
	33	0·0206	0·6815	—0·0261	0·4105
	34	—0·3135	0·7263	0·4555	0·2131
	35	0·4984	0·5139	0·0919	0·5354
	36	0·1244	0·4422	0·2540	0·5040

structure. The 'therapeutic community' and 'relationship oriented' groups seem to constitute more clear clusters than the 'insight oriented' and 'control oriented' groups, which tend to mix with each other.

3.4 PERFORMANCE OF QUANTITATIVE TAXONOMIC METHODS

The comparative performance of the various quantitative taxonomic methods on the treatment environments data base was evaluated according to external criterion validity, internal criterion validity, replicability and inter-rater agreement, as well as across the first three evaluative criteria. The evaluative indices used are described in § 2.3.

Table 3.6 presents the performance and ranking of the quantitative taxonomic methods on treatment environments data, according to external criterion validity. Values on three indices of external criterion validity are presented: percentage of concordance, Cramér's statistic (whose values take into consideration the size of the contingency table and are monotonic with those of other χ^2-related statistics), and a correlation coefficient computed between similarity matrices derived from the cluster configurations produced by a given taxonomic method and the expert. For each index, magnitudes obtained on each set, their average across data sets and derived ranks are presented. Finally, average ranks computed by averaging those for each index, as well as overall ranks, are shown. It can be seen that the ranking obtained according to the three indices are very similar to each other. The best ranked method was complete linkage using correlation coefficients (5.A). Also well ranked were the three *k*-means methods (7.A, 7.C, and 7.B), centroid linkage (6.A) and complete linkage using distance functions (5.C and 5.B). At the other end of the spectrum of external criterion validity, the poorest performing methods were the three single linkage methods (4.B, 4.C and 4.A) and NORMAP/NORMIX (10.X).

Table 3.7 shows the performance and ranking of the quantitative taxonomic methods on treatment environments data according to internal criterion validity. Values for the cophenetic correlation coefficient on data sets A and B and their average, as well as overall ranks, are presented. It can be seen that the best ranked method was centroid linkage (6.a), followed by complete linkage using correlation coefficients (5.A), *k*-means using correlation coefficients (7.A), ordinal multidimensional scaling using Eculidean distances (2.B) and correlation coefficients (2.A), ISODATA (8.X) and complete linkage using distance functions (5.B and 5.C). The poorest ranked methods were single linkage using correlation coefficients (4.A) and the Rubin-Friedman covariance criterion optimization technique (9.X). Next to the poorest performing methods were Q-factor analysis (1.A), NORMAP/NORMIX (10.X) and the Chernoff's faces method (3.X).

Table 3.8 exhibits the performance and ranking of the quantitative taxonomic methods on treatment environments data according to replicability. For each method, both a replicability correlation coefficient

Table 3.6
Performance and rankings of quantitative taxonomic methods according to external criterion validity on the treatment environments data base

	Quantitative taxonomic method	% Concordance				Cramér's statistic				Correlation coefficient (between deriv. simil. matrices)				Avg. rank	Overall rank
		Data set A	Data set B	Avg.	Rank	Data set A	Data set B	Avg.	Rank	Data set A	Data set B	Avg.	Rank		
1.A	Q-factor analysis, corr. coef.	63·89	63·89	63·89	8·5	0·588	0·652	0·620	9	0·272	0·350	0·311	9	8·8	9
2.A	Multidimen. scal., corr. coef.	59·72	61·11	60·42	10	0·586	0·642	0·614	10	0·291	0·316	0·304	10	10	10
2.B	Multidimen. scal., Euclid. dist.	47·23	58·34	52·79	12	0·451	0·586	0·519	13	0·300	0·193	0·247	13	12·7	13
2.C	Multidimen. scal., city-bl. dist.	55·56	38·89	47·23	14	0·619	0·384	0·502	14	0·428	0·058	0·243	14	14	14
3.X	Chernoff's faces	56·95	48·61	52·78	13	0·570	0·512	0·541	12	0·321	0·235	0·278	11	12	12
4.A	Single linkage, corr. coef.	33·33	33·33	33·33	16	0·447	0·344	0·396	16	0·128	0·022	0·075	16	16	16
4.B	Single linkage, Euclid. dist.	30·56	30·56	30·56	17·5	0·294	0·294	0·294	17·5	0·003	0·003	0·003	17·5	17·5	17·5
4.C	Single linkage, city-bl. dist.	30·56	30·56	30·56	17·5	0·294	0·294	0·294	17·5	0·003	0·003	0·003	17·5	17·5	17·5
5.A	Complete linkage, corr. coef.	83·33	77·78	80·56	1	0·796	0·767	0·782	1	0·602	0·551	0·577	1	1	1
5.B	Complete linkage, Euclid. dist.	69·44	66·67	68·06	7	0·658	0·639	0·649	7	0·381	0·417	0·399	6·5	6·8	7
5.C	Complete linkage, city-bl. dist.	80·56	58·33	69·45	6	0·762	0·537	0·650	6	0·542	0·186	0·364	8	6·7	6
6.A	Centroid linkage, corr. coef.	72·22	75·00	73·61	4	0·690	0·726	0·708	5	0·434	0·493	0·464	5	4·7	5
7.A	*k*-Means, corr. coef.	75·00	80·56	77·78	2	0·701	0·781	0·741	2	0·433	0·570	0·502	2	2	2
7.B	*k*-Means, Euclid. dist.	83·33	63·89	73·61	4	0·796	0·642	0·719	4	0·602	0·355	0·479	4	4	4
7.C	*k*-Means, city-bl. dist.	83·33	63·89	73·61	4	0·801	0·656	0·729	3	0·589	0·403	0·496	3	3·3	3
8.X	ISODATA	63·89	63·89	63·89	8·5	0·661	0·615	0·638	8	0·407	0·391	0·399	6·5	7·7	8
9.X	Rubin-Friedman	55·56	52·78	54·17	11	0·502	0·632	0·567	11	0·181	0·357	0·269	12	11·3	11
10.X	NORMAP/NORMIX	50·00	44·44	47·22	15	0·464	0·372	0·418	15	0·178	0·036	0·107	15	15	15

Table 3.7

Performance and rankings of quantitative taxonomic methods according to internal criterion validity on the treatment environments data base

Quantitative taxonomic method	Cophenetic correlation Data set A	Data set B	Average	Rank
1.A Q-factor analysis, corr. coef.	0·265	0·481	0·373	16
2.A Multidimen. scal., corr. coef.	0·664	0·448	0·556	6
2.B Multidimen. scal., euclid dist.	0·628	0·508	0·568	4
2.C Multidimen. scal., city-bl. dist.	0·491	0·457	0·474	11
3.X Chernoff's faces	0·413	0·346	0·380	14
4.A Single linkage, corr. coef.	0·254	0·225	0·240	18
4.B Single linkage, Euclid. dist.	0·320	0·449	0·385	13
4.C Single linkage, city-bl. dist.	0·285	0·549	0·417	12
5.A Complete linkage, corr. coef.	0·623	0·600	0·612	2
5.B Complete linkage, Euclid. dist.	0·603	0·495	0·549	7
5.C Complete linkage, city-bl. dist.	0·577	0·493	0·535	8
6.A Centroid linkage, corr. coef.	0·667	0·677	0·672	1
7.A *k*-Means, corr. coef.	0·542	0·627	0·585	3
7.B *k*-Means, Euclid. dist.	0·568	0·442	0·505	9
7.C *k*-Means, city-bl. dist.	0·542	0·456	0·499	10
8.X ISODATA	0·562	0·551	0·557	5
9.X Rubin-Friedman	0·220	0·360	0·290	17
10.X NORMAP/NORMIX	0·542	0·207	0·375	15

Table 3.8

Performance and rankings of quantitative taxonomic methods according to replicability on treatment environments data

Quantitative taxonomic method	Correlation coefficient (between expert-method cross-classification tables)	Rank
1.A Q-factor analysis, corr. coef.	0·908	5
2.A Multidimen. scal., corr. coef.	0·288	16
2.B Multidimen. scal., Euclid dist.	0·274	17
2.C Multidimen. scal., city-bl. dist.	0·155	18
3.X Chernoff's faces	0·358	15
4.A Single linkage, corr. coef.	0·871	8
4.B Single linkage, Euclid. dist.	1·000	1
4.C Single linkage, city-bl. dist.	0·999	2
5.A Complete linkage, corr. coef.	0·941	4
5.B Complete linkage, Euclid. dist.	0·540	12
5.C Complete linkage, city-bl. dist.	0·650	11
6.A Centroid linkage, corr. coef.	0·968	3
7.A *k*-Means, corr. coef.	0·889	7
7.B *k*-Means, Euclid. dist.	0·818	9
7.C *k*-Means, city-bl. dist.	0·796	10
8.X ISODATA	0·907	6
9.X Rubin-Friedman	0·495	13
10.X NORMAP/NORMIX	0·360	14

computed between corresponding entries of the two expert-method cross-classification tables from data sets A and B and an overall rank, are presented. The top methods were single linkage methods using Euclidean distances (4.B) and city-block distances (4.C), followed by centroid linkage (6.A) and complete linkage using correlation coefficients (5.A). The poorest performing methods were the ordinal multidimensional scaling methods (2.C, 2.B and 2.A) and, next to them, Chernoff's faces (3.X) and NORMAP/NORMIX (10.X).

Table 3.9
Ranking of the quantitative taxonomic methods across all three major evaluative criteria (external criterion validity (ECV), internal criterion validity (ICV) and replicability) on treatment environments data

Quantitative taxonomic method		ECV rank	ICV rank	Replic. rank	Average rank	Overall rank
1.A	Q-factor analysis, corr. coef.	9	16	5	10	9
2.A	Multidimen. scal., corr. coef.	1·0	6	16	10·7	12
2.B	Multidimen. scal., Euclid. dist.	13	4	17	11·3	13
2.C	Multidimen. scal., city-bl. dist.	14	11	18	14·3	17
3.X	Chernoff's faces	12	14	15	13·7	14·5
4.A	Single linkage, corr. coef.	16	18	8	14	16
4.B	Single linkage, Euclid. dist.	17·5	13	1	10·5	10·5
4.C	Single linkage, city-bl. dist.	17·5	12	2	10·5	10·5
5.A	Complete linkage, corr. coef.	1	2	4	2·3	1
5.B	Complete linkage, Euclid. dist.	7	7	12	8·7	8
5.C	Complete linkage, city-bl. dist.	6	8	11	8·3	7
6.A	Centroid linkage, corr. coef.	5	1	3	3	2
7.A	*k*-Means, corr. coef.	2	3	7	4	3
7.B	*k*-Means, Euclid. dist.	4	9	9	7·3	5
7.C	*k*-Means, city-bl. dist.	3	10	10	7·7	6
8.X	ISODATA	8	5	6	6·3	4
9.X	Rubin-Friedman	11	17	13	13·7	14·5
10.X	NORMAP/NORMIX	15	15	14	14·7	18

Table 3.9 presents the overall ranking of the quantitative taxonomic methods obtained by averaging their ranks for the three major evaluation criteria (external criterion validity, internal criterion validity and replicability). The best ranked method turned out to be complete linkage using correlation coefficients (5.A), closely followed by centroid linkage (6.A) and *k*-means using correlation coefficients (7.A). The overall poorest ranked methods were NORMAP/NORMIX (10.X), ordinal multidimensional scaling using city-block distances (2.C) and single linkage using correlation coefficients (4.A).

Table 3.10

Performance and ranking of the ordinal multidimensional scaling and Chernoff's faces methods according to inter-rater agreement on the treatment environments data base

Quantitative taxonomic method	% Concordance				Cramér's statistic				Correlation coefficient				Avg. rank	Overall rank
	Data set A	Data set B	Avg.	Rank	Data set A	Data set B	Avg.	Rank	Data set A	Data set B	Avg.	Rank		
2.A Multidimen. scal., corr. coef.	83·33	77·78	80·56	2	0·725	0·829	0·777	2	0·753	0·567	0·660	1	1·7	2
2.B Multidimen. scal., Euclid. dist.	94·44	58·33	76·39	3	0·816	0·674	0·745	3	0·920	0·278	0·599	3	3	3
2.C Multidimen. scal., city-bl. dist.	100·0	63·89	81·94	1	1·000	0·674	0·837	1	1·000	0·302	0·651	2	2	1
3.X Chernoff's faces	50·00	38·89	44·45	4	0·373	0·360	0·367	4	0·132	0·132	0·132	4	4	4

Table 3.11

Performance and ranking of relationship measures according to external criterion validity, internal criterion validity (ICV) and replicability on treatment environments data

	Relationship measure	External criterion validity				ICV		Replic.		Average rank	Overall rank
		% Concordance	Cramér	Corr. coef.	Overall average rank	Cophenetic corr.	Rank	Correl. coef.	Rank		
A	Correlation coefficient	63·02	0·633	0·365	1	0·498	2	0·747	1	1·33	1
B	Euclidean distance	56·26	0·545	0·282	2	0·502	1	0·658	2	1·67	2
C	City-block distance	55·21	0·544	0·277	3	0·481	3	0·650	3	3	3

Table 3.10 exhibits the performance and ranking of the ordinal multidimensional scaling and Chernoff's faces methods on treatment environments data according to inter-rater agreement (these taxonomic methods require the participation of human judges to complete the clustering process). The evaluation indices are formally similar to those used for external criterion validity. The Chernoff's faces method (3.X) clearly had a poorer ranking than the ordinal multidimensional scaling methods (2.A, 2.B and 2.C) which had much smaller differences among them.

3.5 PERFORMANCE OF RELATIONSHIP MEASURES

The performance of the three relationship measures used in the present study, namely correlation coefficient, Euclidean distance and city-block distance, was assessed by averaging the performance of the corresponding forms of the ordinal multidimensional scaling, single linkage, complete linkage and *k*-means methods, all of which used the three relationship measures. For example, the performance of the correlation coefficient (A) was assessed by averaging the performances of 2.A, 4.A, 5.A and 7.A.

Table 3.11 presents the performance and ranking of the relationship measures on treatment environments data according to external criterion validity, internal criterion validity, replicability and across all these three evaluation criteria. The ranks shown should be interpreted with caution, since the actual performance of the relationship measures appear to be quite comparable to each other.

3.6 CLUSTER-BY-CLUSTER AGREEMENT BETWEEN EXPERT AND TAXONOMIC METHODS

The classification produced by each clustering method on the treatment environments data base was cross-tabulated with the expert-proposed classification. The resulting 44 cross-classification tables (22 for each data set, including two tables for each of the three ordinal multidimensional scaling methods and the Chernoff's faces method as two judges produced separate clusterings for each of these methods), were summed cell-by-cell. Table 3.12 presents this summary information.

Table 3.13 was derived from Table 3.12 to present the percentages of each one of the expert-proposed groups classified into the various clusters produced by the quantitative taxonomic methods. The highest levels of overall agreement between the expert and the taxonomic methods were found for 'therapeutic community' (67·17%) and 'relationship oriented'

Table 3.12
Cell-by-cell sum of cross-classification tables between expert and quantitative taxonomic methods on the treatment environments data base

Expert-produced Groups	Clusters produced by quantitative taxonomic methods				
	Therapeutic community	Relationship oriented	Insight oriented	Control oriented	
Therapeutic community	266	82	18	30	396
Relationship oriented	90	260	17	29	396
Insight oriented	80	70	148	98	396
Control oriented	15	73	70	238	396
	451	485	253	395	1584

(65·66%), somewhat closely followed by 'control oriented' (60·10%). The agreement level for 'insight oriented' was much lower (37·37%). This finding probably reflects the less defined conceptual identification of the 'insight oriented' category.

Table 3.13
Percentages of each one of the expert-produced groups classified into the various clusters produced by the quantitative taxonomic methods on the treatment environments data base

Expert-produced groups	Clusters produced by quantitative taxonomic methods				
	Therapeutic community	Relationship oriented	Insight oriented	Control oriented	
Therapeutic community	67·17	20·71	4·54	7·58	100·00
Relationship oriented	22·73	65·66	4·29	7·32	100·00
Insight oriented	20·20	17·68	37·37	24·75	100·00
Control oriented	3·79	18·43	17·68	60·10	100·00

Regarding areas of misclassification, it can not noted in Table 3.12 that the main confusion associated with the expert-based groups 'therapeutic community' and 'relationship oriented' occurred with each other, while confusions associated with the last two groups tended to occur with each other in addition to considerable confusion with the first two groups. This finding is in line with the description of the four types of treatment environments reported by Price and Moos (1975), which indicate that the 'therapeutic community' and 'relationship oriented' types are similar to each other in that both have relatively small numbers of patients and low patient/staff ratios, while both the 'insight oriented' and the 'control oriented' types have relatively large numbers of patients and high patient/staff ratios.

The relatively unclear cluster structure noted on this data set is in line with the early developmental stage of social ecology as a science. However, future conceptual developments and further empirical observations using clustering and other exploratory data analyses should promote the clarification of basic relationships and the elucidation of useful patterns in this tremendously interesting area of behavioral science.

4

Cluster Analysis of Archetypal Psychiatric Patients

4.1 INTRODUCTION

Along with the increasing awareness of the importance of psychiatric diagnosis for etiological research and the study and optimization of clinical decisions, significant criticism of the structure and content of current diagnostic systems has been formulated on scientific grounds (Strauss, 1973; Panzetta, 1974). One of the critical needs is to obtain more reliable and valid groupings of psychiatric patients. Patient groupings are not only central to typological or categorical diagnostic models, such as the current International Classification of Diseases (ICD-9), World Health Organization, 1978), but are also important components of the structurally innovative multiaxial models (Mezzich, 1979, 1980), as currently conceptualized for psychiatric disorders in adults (e.g. Strauss, 1975; American Psychiatric Association, 1980) and children (Rutter, Shaffer and Shepherd, 1975).

Research findings regarding the clinical predictive value and biosocial correlates of some diagnostic categories have been used in the preparation of standard diagnostic systems, but the major basis of these seems to have been consensual validity. This refers to the judgmental agreement of experts on a committee, frequently with input from other experienced clinicians, on the appropriateness of a proposed diagnostic system. This illustrates the pertinence of the structural study of diagnostic concepts. Such structural study may involve the elucidation of conceptual norms in terms of mean profiles of groups of archetypal patients representing particular diagnostic types, and the assessment of the cohesiveness or homogeneity and the differentiability of diagnostic categories.

An interesting perspective in the study of diagnostic categories can be offered by the coordinated use of quantitative taxonomic methods. The multiplicity of such methods available today, representing quite different approaches to the taxonomic problem, provide the possibility of studying the structure of diagnostic concepts and groups through the analysis of patterns of agreement of quantitative taxonomic methods with experienced clinicians.

A set of archetypal psychiatric patients developed by experienced psychiatrists to represent four diagnostic categories through ratings on 17 psychopathological variables were used in this study. The various quantitative taxonomic methods described in Chapter 2 were applied to this data base. Some of the clustering processes and resulting configurations are presented for illustrative purposes. The comparative performance of the taxonomic methods and relationship measures on this data base is then presented and discussed according to external criterion validity, internal criterion validity, replicability and across these three evaluative criteria as well as according to inter-rater reliability. Also, the normative profiles of the four groups of archetypal psychiatric patients are presented in addition to the levels of agreement between the clinicians and the taxonomic methods for each group.

4.2 DATA BASE

The data base for this study consisted of 88 archetypal psychiatric patients fabricated by 22 experienced psychiatrists on the academic and clinical faculty of the Stanford University Department of Psychiatry and Behavioral Sciences. Each psychiatrist was invited to think, one at a time, on a typical patient for each one of four diagnostic categories: manic-depressive depressed, manic-depressive manic, simple schizophrenic and paranoid schizophrenic. These four diagnostic categories are part of the nomenclature of mental disorders (DSM-II) issued in 1968 by the American Psychiatric Association. The DSM-II descriptions of these categories are as follows:

(1) *Manic-depressive illness, depressed.* This disorder is characterized by severely depressed mood and by mental and motor retardation progressing occasionally to stupor. Uneasiness, apprehension, perplexity and agitation may also be present. When illusions, hallucinations and delusions (usually of guilt or of hypochondriacal or paranoid ideas) occur, they are attributable to the dominant mood disorder.

(2) *Manic-depressive illness, manic.* This disorder is characterized by excessive elation, irritability, talkativeness, flight of ideas, and accelerated speech and motor activity.

(3) *Simple schizophrenia.* This psychosis is characterized chiefly by a slow and insidious reduction of external attachments and interests and by apathy and indifference leading to impoverishment of interpersonal relations, mental deterioration and adjustment on a lower level of functioning. In general, the condition is less dramatically psychotic than are the hebephrenic, catatonic and paranoid types of schizophrenia.

(4) *Paranoid schizophrenia.* This type of schizophrenia is characterized primarily by the presence of persecutory or grandiose delusions, often associated with hallucinations. Excessive religiosity is sometimes seen. The patient's attitude is frequently hostile and aggressive and his behavior tends to be consistent with his delusions.

Each archetypal patient was characterized by 0-6 severity ratings on 17 items (the 16 original items plus one added recently: 'excitement') from the Brief Psychiatric Rating Scale (BPRS) (Overall and Gorham, 1962). These ratings were recorded on a BPRS blank labeled at the top with the diagnostic category the patient was supposed to represent. The BPRS items used in this study are the following:

(1) *Somatic concern.* Preoccupation with physical health, fear of physical illness, hypochondriasis.
(2) *Anxiety.* Worry, fear, overconcern for present or future.
(3) *Emotional withdrawal.* Lack of spontaneous interaction, isolation, deficiency in relating to others.
(4) *Conceptual disorganization.* Thought processes confused, disconnected, disorganized, disrupted.
(5) *Guilt Feelings.* Self-blame, shame, remorse for past behavior.
(6) *Tension.* Physical and motor manifestations or nervousness, overactivity, tension.
(7) *Mannerisms and posturing.* Peculiar, bizarre, unnatural motor behavior (not including tic).
(8) *Grandiosity.* Exaggerated self-opinion, arrogance, conviction of unusual power or abilities.
(9) *Depressive mood.* Sorrow, sadness, despondency, pessimism.
(10) *Hostility.* Animosity, contempt, belligerence, disdain for others.
(11) *Suspiciousness.* Mistrust, belief others harbor malicious or discriminatory intent.
(12) *Hallucinatory behavior.* Perceptions without normal external stimulus correspondence.
(13) *Motor retardation.* Slowed weakened movements or speech, reduced body tone.
(14) *Uncooperativeness.* Resistance, guardedness, rejection of authority.

Table 4.1
Data matrix of archetypal psychiatric patients measured on 17 psychopathological items from the Brief Psychiatric Rating Scale: data set A

Archetypal patients		Psychopathological symptoms/signs																
		1	2	3	4	5	6	7	8	9	10	11	12	13	14	15	16	17
Manic-depressive depressed	1	4	3	3	0	4	3	0	0	6	3	2	0	5	2	2	2	1
	2	5	5	6	2	6	1	0	0	6	1	0	1	6	4	1	4	0
	3	6	5	6	5	6	3	2	0	6	0	5	3	6	5	5	0	0
	4	5	5	1	0	6	1	0	0	6	0	1	2	6	0	3	0	2
	5	6	6	5	0	6	0	0	0	6	0	4	3	5	3	2	0	0
	6	3	3	5	1	4	2	1	0	6	2	1	1	5	2	2	1	1
	7	5	5	5	2	5	4	1	1	6	2	3	0	6	3	5	2	3
	8	4	5	5	1	6	1	1	0	6	1	1	0	5	2	1	1	0
	9	5	3	5	1	6	3	1	0	6	2	1	1	6	2	5	5	0
	10	3	5	5	3	2	4	2	0	6	3	2	0	6	1	4	5	1
	11	5	6	6	4	6	3	1	0	6	2	0	0	6	4	4	6	0
Manic-depressive manic	12	2	2	1	2	0	3	1	6	2	3	3	2	1	4	4	0	6
	13	0	0	0	4	1	5	0	6	0	5	4	4	0	5	5	0	6
	14	0	3	0	5	0	6	0	6	0	3	2	0	0	3	4	0	6
	15	0	0	0	3	0	6	0	6	1	3	1	1	0	2	3	0	6
	16	3	4	0	0	0	5	0	6	0	6	0	0	0	5	0	0	6
	17	2	4	0	3	1	5	1	6	2	5	3	0	0	5	3	0	6
	18	1	2	0	2	1	4	1	5	1	5	1	1	0	4	1	0	6
	19	0	2	0	2	1	5	1	5	0	2	1	1	0	3	1	0	6
	20	0	0	0	6	0	5	1	6	0	5	5	4	0	5	6	0	6
	21	5	5	1	4	0	5	5	6	0	4	4	3	0	5	5	0	6
	22	1	3	0	4	1	4	2	6	3	3	2	0	0	4	3	0	6
Simple schizophrenic	23	3	2	5	2	0	2	2	1	2	1	2	0	1	2	2	4	0
	24	4	4	5	4	3	3	1	0	4	2	3	0	3	2	4	5	0
	25	2	0	6	3	0	0	5	0	0	3	3	2	3	5	3	6	0
	26	1	1	6	2	0	0	1	0	0	3	0	1	0	1	1	6	0
	27	3	3	5	6	3	2	5	0	3	0	2	5	3	3	5	6	2
	28	3	0	5	4	0	0	3	0	2	1	1	1	2	3	3	6	0
	29	3	3	5	4	2	4	2	1	3	1	1	1	4	2	2	5	2
	30	3	2	5	2	2	2	2	1	2	2	3	1	2	2	3	5	0
	31	3	3	6	6	1	3	5	1	3	2	2	5	3	3	6	6	1
	32	1	1	5	3	1	1	3	0	1	1	1	0	5	1	2	6	0
	33	2	3	5	4	2	3	0	0	3	2	2	0	0	2	4	5	0
Paranoid schizophrenic	34	2	4	3	5	0	3	1	4	2	5	6	5	0	5	6	3	3
	35	2	4	1	1	0	3	1	6	0	6	6	4	0	6	5	0	4
	36	5	5	5	6	0	5	5	6	2	5	6	6	0	5	6	0	2
	37	1	4	2	1	1	1	0	5	1	5	6	5	0	6	6	0	1
	38	4	5	6	3	1	6	3	5	2	6	6	4	0	5	6	0	5
	39	4	5	4	6	2	4	2	4	1	5	6	5	1	5	6	2	4
	40	3	4	3	4	1	5	2	5	2	5	5	3	1	5	5	1	5
	41	2	5	4	3	1	4	3	4	2	5	5	4	0	5	4	1	4
	42	3	3	4	4	1	5	5	5	0	5	6	5	1	5	5	3	4
	43	4	4	2	6	1	4	1	5	3	5	6	5	1	5	6	2	4
	44	3	5	5	5	2	5	4	5	2	4	6	5	0	5	6	5	5

(15) *Unusual thought content.* Unusual, odd, strange, bizarre thought content.
(16) *Blunted effect.* Reduced emotional tone, reduction in normal intensity of feelings, flatness.
(17) *Excitement.* Heightened emotional tone, agitation, increased reactivity.

In order to study the replicability of resulting cluster configurations, each diagnostic group of 22 patients was randomly divided into two groups of 11 each, to constitute collectively data sets A and B, each of 44 archetypal patients. Tables 4.1 and 4.2 present data matrices constituted by the 17 BPRS ratings for each of the 44 archetypal patients of data sets A and B, respectively.

Table 4.3 presents, separately for data sets A and B, the means and standard deviations of the four diagnostic groups of archetypal psychiatric patients, on 17 BPRS items. Figure 4.1 exhibits the four group mean profiles, for each data set. It can be seen that the mean values of corresponding diagnostic groups on both data sets are closely similar, as would be expected from the fact that data sets A and B were produced by randomly halving, by diagnostic group, the original set of 88 archetypal patients. Inspection of the mean values of a given variable across diagnostic groups provides an indication of its differentiating value. For example, 'suspiciousness' (variable 11), whose mean is clearly highest for paranoid schizophrenia, seems to differentiate well this group from the other three. However, the standard deviations on the various diagnostic groups of the variable under examination have to be taken into consideration for a more precise evaluation of its differentiating power.

Table 4.2

Data matrix of archetypal psychiatric patients measured on 17 psychopathological items from the Brief Psychiatric Rating Scale: data set B

Archetypal patients		Psychopathological symptoms/signs 1	2	3	4	5	6	7	8	9	10	11	12	13	14	15	16	17
Manic-depressive depressed	1	2	2	5	2	5	1	0	0	5	2	2	0	5	2	3	4	0
	2	4	4	1	1	3	3	0	0	5	2	2	2	5	1	1	0	0
	3	6	5	5	4	6	4	0	1	6	3	2	3	6	2	4	4	0
	4	4	2	5	2	5	4	0	0	6	2	2	0	5	2	3	1	0
	5	5	4	3	2	6	2	1	0	6	2	1	1	5	2	3	1	0
	6	5	4	6	4	6	3	0	0	6	2	0	2	6	4	4	2	0
	7	5	5	2	4	5	5	1	0	6	1	4	4	6	4	5	2	0
	8	6	5	5	2	6	3	1	0	6	2	1	1	5	0	1	0	1
	9	5	5	5	2	5	3	0	0	6	1	2	2	5	3	4	0	3
	10	4	3	4	3	4	3	0	0	6	2	2	4	5	2	4	0	0
	11	3	2	4	2	5	2	1	0	5	3	1	3	5	2	2	1	2
Manic-depressive manic	12	2	2	0	4	0	5	0	6	2	5	5	2	0	5	3	0	6
	13	3	1	0	2	0	2	1	3	1	4	3	2	0	3	1	0	5
	14	1	0	0	5	0	5	2	6	3	5	3	3	0	4	2	2	6
	15	1	1	0	4	1	3	2	6	1	3	3	1	0	4	3	0	6
	16	2	1	0	2	2	3	1	5	1	5	2	1	0	4	1	0	6
	17	0	0	0	4	0	5	0	6	0	5	3	0	0	3	2	1	6
	18	3	5	0	4	0	5	2	6	0	5	4	4	0	5	5	0	6
	19	1	0	0	4	2	4	2	6	2	3	2	1	0	3	3	2	6
	20	2	1	0	3	1	4	0	4	2	5	3	2	0	6	3	0	6
	21	0	1	0	4	0	3	0	5	2	3	3	4	0	3	4	0	5
	22	1	2	0	3	1	4	1	5	2	3	1	2	0	2	2	0	5
Simple schizophrenic	23	3	1	5	5	3	1	3	2	3	3	4	2	3	5	5	5	2
	24	1	1	4	2	0	1	2	1	0	1	1	0	1	2	2	4	0
	25	2	1	6	2	1	1	5	1	4	2	1	4	5	4	5	6	0
	26	2	2	5	4	1	2	2	1	1	2	3	3	3	3	4	6	0
	27	3	2	6	4	2	1	1	0	2	2	1	2	2	2	3	5	1
	28	0	0	6	5	0	0	4	0	1	1	0	0	4	2	3	6	0
	29	3	3	6	5	2	3	2	2	2	2	3	2	2	1	4	5	0
	30	2	2	5	2	1	1	2	1	2	2	2	2	1	3	4	5	2
	31	2	2	6	4	2	2	3	2	3	1	3	3	4	3	3	5	0
	32	1	2	5	3	1	2	0	0	1	2	1	0	0	2	3	5	0
	33	0	0	6	1	0	2	2	0	0	0	2	0	2	2	1	5	1
Paranoid schizophrenic	34	3	5	4	6	2	5	5	6	2	5	6	6	1	6	5	2	5
	35	3	4	2	2	0	3	2	4	1	4	5	3	0	5	3	2	4
	36	5	6	6	2	0	3	3	6	2	5	6	3	0	5	4	3	2
	37	3	5	0	5	2	6	3	6	2	5	6	4	0	6	5	0	3
	38	4	3	3	5	1	3	1	5	1	5	6	4	1	5	6	1	3
	39	1	4	4	3	0	3	0	6	0	5	6	3	0	6	5	3	1
	40	2	5	5	5	2	4	4	4	2	5	6	5	1	4	5	4	4
	41	2	4	3	6	4	4	3	5	2	5	5	5	1	4	6	2	4
	42	2	3	3	3	0	3	2	3	1	5	6	3	1	6	4	3	3
	43	1	3	4	4	2	2	0	5	2	4	5	4	0	3	4	2	1
	44	1	3	2	3	0	2	1	5	0	5	6	3	0	4	3	2	3

Table 4.3

Means and standard deviations of four diagnostic groups, manic-depressive depressed (MDD), manic-depressive manic (MDM), simple schizophrenic (SS) and paranoid schizophrenic (PS), on 17 Brief Psychiatric Rating Scale items: data sets A and B

Diagnostic groups	Statistic	Somatic concern 1	Anxiety 2	Emotional withdrawal 3	Conceptual disorganization 4	Guilt feelings 5	Tension 6	Mannerisms and posturing 7	Grandiosity 8	Depressive mood 9	Hostility 10	Suspiciousness 11	Hallucinatory behavior 12	Motor retardation 13	Uncooperativeness 14	Unusual thought content 15	Blunted effect 16	Excitement 17
Data set A																		
MDD	Mean	4·64	4·64	4·73	1·73	5·18	2·27	0·82	0·09	6·00	1·45	1·82	1·00	5·64	2·54	3·09	2·36	0·73
($N=11$)	SD	1·03	1·12	1·49	1·68	1·33	1·35	0·75	0·30	0·00	1·13	1·60	1·18	0·51	1·44	1·58	2·25	1·01
MDM	Mean	1·27	2·27	0·18	3·18	0·46	4·82	1·09	5·82	0·82	4·00	2·36	1·46	0·09	4·09	3·18	0·00	6·00
($N=11$)	SD	1·62	1·74	0·40	1·66	0·52	0·37	1·45	0·40	1·08	1·26	1·58	1·57	0·30	1·04	1·88	0·00	0·00
SS	Mean	2·54	2·00	5·27	3·64	1·27	1·82	2·64	0·36	2·09	1·64	1·82	1·46	2·36	2·36	3·18	5·46	0·46
($N=11$)	SD	0·93	1·34	0·47	1·43	1·19	1·40	1·75	0·50	1·30	0·92	0·98	1·86	1·57	1·12	1·47	0·69	0·82
PS	Mean	3·00	4·36	3·54	4·00	0·91	4·09	2·46	4·91	1·54	5·09	5·82	4·64	0·36	5·18	5·54	1·54	3·73
($N=11$)	SD	1·18	0·67	1·51	1·84	0·70	1·38	1·70	0·70	0·93	0·54	0·40	0·81	0·50	0·40	0·69	1·64	1·27
Data set B																		
MDD	Mean	4·46	3·73	4·09	2·54	5·09	3·00	0·36	0·09	5·73	2·00	1·73	2·00	5·27	2·18	3·09	1·36	0·54
($N=11$)	SD	1·21	1·27	1·51	1·04	0·94	1·10	0·50	0·30	0·47	0·63	1·01	1·41	0·47	1·17	1·30	1·50	1·04
MDM	Mean	1·46	1·27	0·00	3·54	0·64	3·91	1·00	5·27	1·46	4·18	2·91	2·00	0·00	3·82	2·64	0·46	5·73
($N=11$)	SD	1·04	1·42	0·00	0·93	0·81	1·04	0·89	1·01	0·93	0·98	1·04	1·26	0·00	1·17	1·21	0·82	0·47
SS	Mean	1·73	1·46	5·46	3·64	1·18	1·46	2·36	0·91	1·73	1·64	1·91	1·64	2·46	2·64	3·36	5·18	0·54
($N=11$)	SD	1·10	0·93	0·69	1·43	0·98	0·82	1·36	0·83	1·27	0·81	1·22	1·43	1·51	1·12	1·21	0·60	0·82
PS	Mean	2·46	4·09	3·27	4·00	1·18	3·46	2·18	5·00	1·36	4·82	5·73	3·91	0·46	4·91	4·54	2·18	3·00
($N=11$)	SD	1·29	1·04	1·62	1·48	1·33	1·21	1·60	1·00	0·81	0·40	0·47	1·04	0·52	1·04	1·04	1·08	1·26

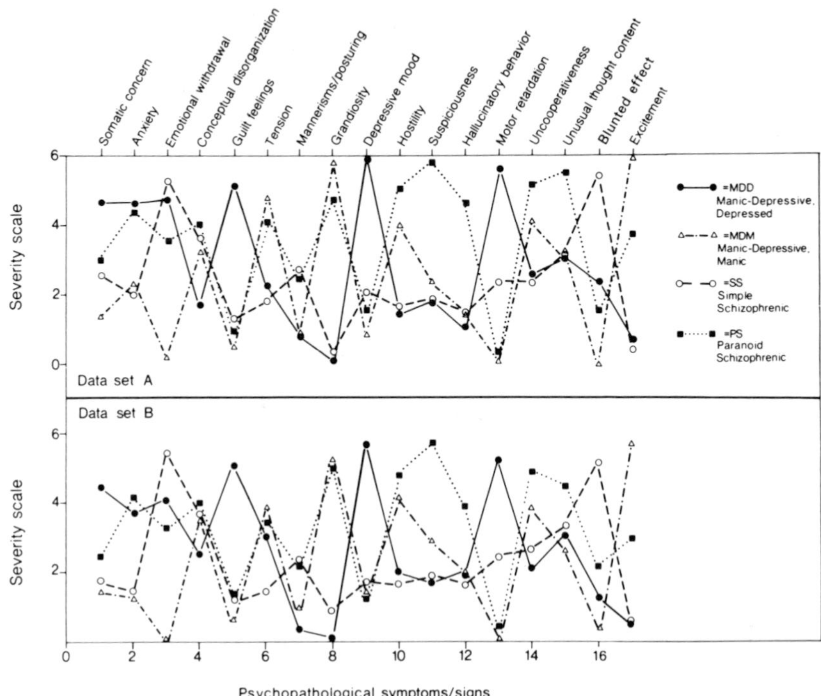

Fig. 4.1 Profiles of four groups of archetypal psychiatric patients of data sets A and B.

4.3 ILLUSTRATIVE CLUSTER ANALYTIC RESULTS

The quantitative taxonomic methods listed in Table 2.1 were used on both data sets A and B of archetypal psychiatric patients. Four figures and two tables will be presented in this section to show the clustering processes and results obtained from the application of some of these taxonomic methods. Our idea is to provide examples of specific clustering outcomes obtained on archetypal psychiatric patients, while illustrating the application of some of these taxonomic methods, which are not necessarily the best methods for this data base.

Two of the taxonomic methods to be presented here for both data sets A and B are also presented, for comparative purposes, in the chapters corresponding to the other three data bases. One of these two methods is the complete linkage hierarchical approach, which was the best ranked taxonomic method across evaluative criteria and data bases, as described in Chapter 7. The other method is ordinal multidimensional scaling, also using correlation coefficients, which is particularly useful for visualizing data structure. Additionally, the results obtained on data set A with the ISODATA method, which ranked in second place for archetypal psychiatric patients, and those obtained on data set B with the Rubin-Friedman reallocation technique, will be presented.

Figures 4.2 and 4.3 show two-dimensional representations of the archetypal psychiatric patients of data sets A and B, respectively, obtained by using ordinal multidimensional scaling and correlation coefficients. Each archetypal patient is represented by a symbol indicating its diagnostic group identification according to the experienced psychiatrists who developed those patients. The two plots are somewhat similar to each other. In both of

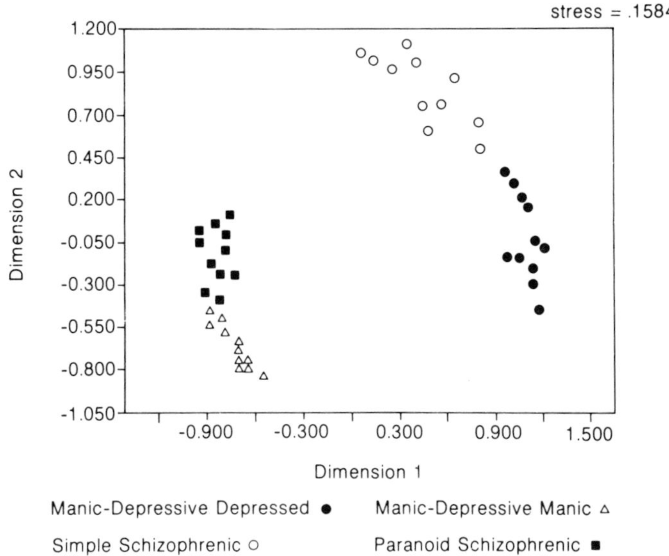

Fig. 4.2 Ordinal multidimensional scaling representation, using correlation coefficients, of the 44 archetypal psychiatric patients of data set A.

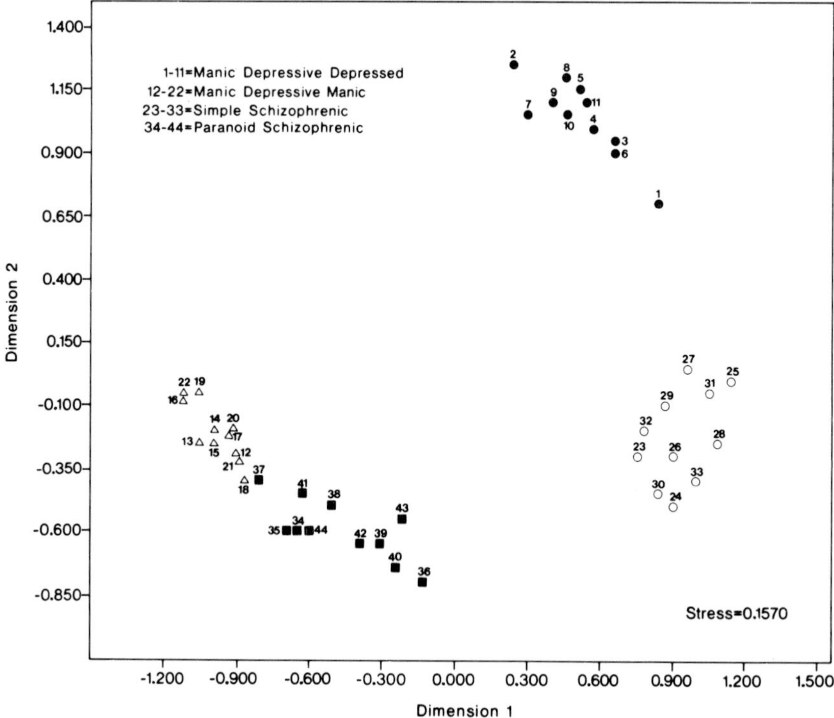

Fig. 4.3 Ordinal multidimensional scaling representation, using correlation coefficients, of the 44 archetypal psychiatric patients of data set B.

them, but especially clearly in that for data set B, the groups of depressed and simple schizophrenic patients are rather tightly clustered as such and easily distinguishable from each other. Both groups are quite distant from the manic and the paranoid schizophrenic groups, which in both plots are located in the lower left corner of the field. These two groups have most of their members clumping together, but appear adjacent to each other rather than constituting clearly separated clusters. Two judges were asked to inspect visually the plots containing unlabeled points, and to group them into four clusters. For data set A, one of the judges correctly clustered the depressed and simple schizophrenic archetypal patients, put together seven manic patients in one cluster and lumped the other four into the same cluster with the 11 paranoid schizophrenic patients. The other judge correctly clustered the 11 simple schizophrenic patients, put five and six of the depressed patients into two separate clusters, and lumped all the manic and paranoid schizophrenic patients into one cluster. For data set B, both judges correctly clustered the 44 archetypal patients into four groups, except one paranoid schizophrenic patient (No. 37) who was clustered with the 11 manics.

Figures 4.4 and 4.5 exhibit dendrograms representing the complete linkage agglomerative cluster analysis using correlation coefficients as relationship measures, of the archetypal patients of data sets A and B, respectively. In each dendrogram, the whole hierarchical clustering process is depicted, from the beginning at the top where each patient is a single cluster, to the end where all patients are lumped into one cluster. A four-cluster solution was obtained at similarity level 0·284 in data set A, and at similarity level 0·276 in data set B. The cluster memberships obtained were quite consistent with the groupings created by the clinicians. The only exceptions were patient No. 21 in data set A and patient No. 18 in data set B, both of them manic archetypal patients, who in each case turned out to be clustered with the paranoid schizophrenics.

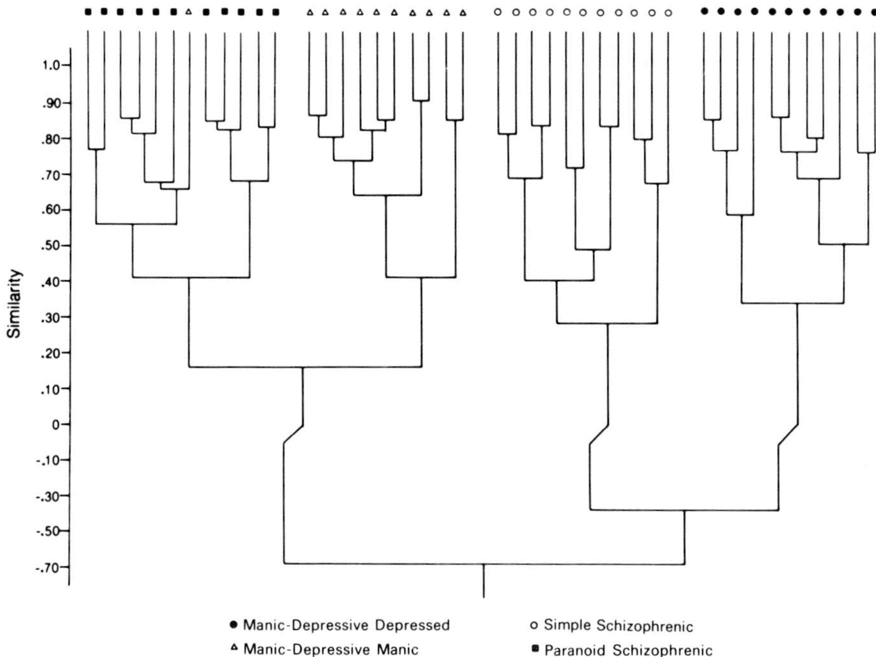

Fig. 4.4 Dendrogram representing a complete linkage cluster analysis, using correlation coefficients, of the 44 archetypal psychiatric patients of data set A.

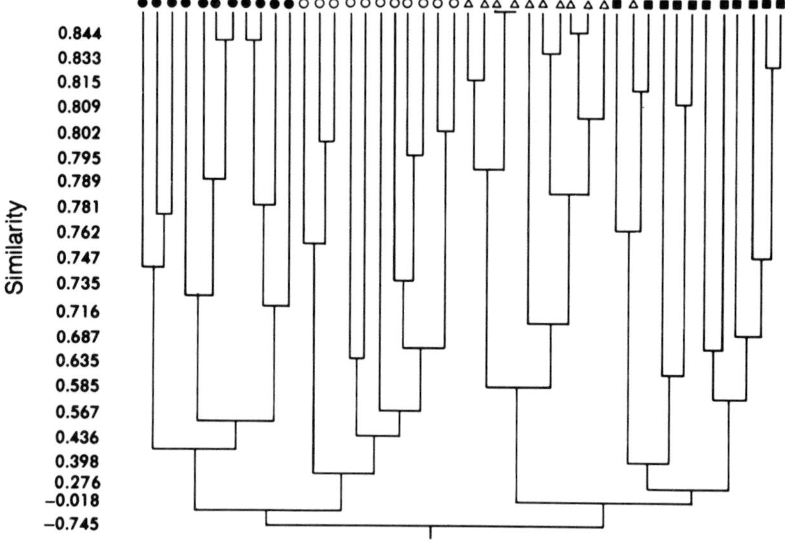

Fig. 4.5 Dendrogram representing a complete linkage cluster analysis, using correlation coefficients, of the 44 archetypal psychiatric patients of data set B.

Table 4.4
ISODATA cluster analysis of the 44 archetypal psychiatric patients of data set A

Iteration 1

Initial cluster membership:

Cluster 1 contains entities:	24	27	29	31	33				
Cluster 2 contains entities:	2	6	9	10	11				
Cluster 3 contains entities:	3	7							
Cluster 4 contains entities:	1	4	5	8					
Cluster 5 contains entities:	12	13	14	15	20	22			
Cluster 6 contains entities:	16	17	18	19					
Cluster 7 contains entities:	21	36	38	39	40	41	42	43	44
Cluster 8 contains entities:	23	25	26	28	30	32			
Cluster 9 contains entities:	34	35	37						

Number of clusters at end of iteration:	9
Average distance of entities from their centers:	3·838
Average distance between cluster centers:	10·871

Table 4.4 cont.

Iteration 2
With: Maximum number of cluster pairs to lump: 2
Minimum distance between clusters to be lumped: 0·10 E 7
Clusters 5 and 6 were lumped together
Clusters 2 and 4 were lumped together

Number of clusters at end of iteration:	7
Average distance of entities from their centers:	4·172
Average distance between cluster centers:	10·622

Iteration 3
With: Maximum number of cluster pairs to lump: 2
Minimum distance between clusters to be lumped: 0·10 E7
New clusters 1 and 6 were lumped together
New clusters 5 and 7 were lumped together

Number of clusters at end of iteration:	5
Average distance of entities from their centers:	4·554
Average distance between cluster centers:	10·948

Iteration 4
With: Maximum number of cluster pairs to lump: 2
Minimum distance between clusters to be lumped: 0·10 E 7
New clusters 2 and 3 were lumped together
New clusters 4 and 5 were lumped together

Number of clusters at end of iteration:	3
Average distance of entities from their centers:	5·469
Average distance between cluster centers:	10·731

Iteration 5
With: Maximum distance between clusters to be split: 1·0
New cluster 2 was split into clusters 2 and 4
New cluster 3 was split into clusters 3 and 5
New cluster 1 was split into clusters 1 and 6

Number of clusters at end of iteration:	6
Average distance of entities from their centers:	4·305
Average distance between cluster centers:	10·689

Iteration 6
With: Maximum number of cluster pairs to be lumped: 2
Minimum distance between clusters to be lumped: 0·10 E 7
New cluster 2 and 4 were lumped together
New cluster 1 and 6 were lumped together

Number of clusters at end of iteration:	4
Average distance of entities from their centers:	4·677
Average distance between cluster centers:	11·146

Final cluster membership
Cluster 1 contains entities: 23-33
Cluster 2 contains entities: 1-11
Cluster 3 contains entities: 12-20, 22
Cluster 4 contains entities: 21, 34-44

Table 4.4. presents the process involved in the ISODATA cluster analysis of the 44 archetypal psychiatric patients of data set A. Through six iterations, various phases of cluster lumping and splitting took place. In iteration 4, the depressed (patients 1-11) and simple schizophrenic (patients 23-33) groups were formed as such, while a third cluster included all manic and paranoid schizophrenic patients. Later in the process, the depressed and simple schizophrenic groups split first and then reconstituted as before. They never mixed with each other or with the other two groups. In iteration 5, the cluster which included all manic and paranoid schizophrenic patients split into one cluster composed of all paranoid schizophrenics plus one manic (No. 21), and another cluster composed of the remaining ten manic patients. It should be noted that archetypal patient No. 21 was already mixed with paranoid schizophrenic patients (eight of them) in iteration 1, which involved a series of settling iterations of *k*-means type. This is

Table 4.5

Rubin-Friedman reallocation technique using maximization of log|**T**|/|**W**| as criterion, applied to the 44 archetypal psychiatric patients of data set B

A series of 'hill-climbing' passes, each consisting in sifting through partitions obtained by moving, one by one, an entity to a different group and retaining the new partition if it significantly improved the criterion value, culminated in the following 'single-move local maximum'

Local maximum No. 1

Members of cluster 1:	18	34	35	36	37	38	39	40	41	42	43	44
Members of cluster 2:	1	2	3	4	5	6	7	8	9	10	11	
Members of cluster 3:	12	13	14	15	16	17	19	20	21	22		
Members of cluster 4:	23	24	25	26	27	28	29	30	31	32	33	

Criterion value: 8·4238

A series of forcing passes each consisting in forcing each entity out of its group and into the group with highest attraction for it, culminated in the following local maximum

Local maximum No. 2

Members of cluster 1:	12	13	14	15	16	17	19	21	22			
Members of cluster 2:	23	24	25	26	27	28	29	30	31	32	33	
Members of cluster 3:	18	34	35	36	37	38	39	40	41	42	43	44
Members of cluster 4:	1	2	3	4	5	6	7	8	9	10	11	

Criterion value: 8·4240

This change in value from the previous local maximum is less than 0·0084239 and is therefore taken as no change.
Therefore, the final group memberships are those corresponding to local maximum No. 1.
Final log|**T**|/|**W**| = 8·4238.

consistent with the outcome of the straight k-means cluster analysis of this data set, which recovered all four diagnostic groups, except that manic patient No. 21 was included in the paranoid schizophrenic cluster.

Table 4.5 shows the results of applying the Rubin-Friedman reallocation technique with maximization of $\log|\mathbf{T}|/|\mathbf{W}|$ as criterion, to the 44 archetypal psychiatric patients of data set B. A series of hill-climbing passes culminated in a cluster configuration which recovered the depressed and simple schizophrenic groups perfectly, included one of the manic patients (No. 18) in the same cluster as the 11 paranoid schizophrenic patients and grouped the remaining ten manic patients into one cluster. The value of the criterion at this point was 8·4238. A series of forcing passes produced a second local maximum. The resulting four clusters had the same membership as those obtained at local maximum No. 1. The criterion value was comparable to that achieved at local maximum No. 1, which led to accepting local maximum No. 1 cluster configuration as the final one.

4.4 PERFORMANCE OF QUANTITATIVE TAXONOMIC METHODS

A comparison of the performance of the various quantitative taxonomic methods was carried out on the archetypal psychiatric patient data base. The evaluative criteria included external criterion validity, internal criterion validity, replicability and inter-rater agreement. The indices employed are described in § 2.3.

Table 4.6 presents the performance and ranking of the quantitative taxonomic methods according to external criterion validity. Concordance percentage, Cramér's statistic and correlation coefficient values were computed for each taxonomic method separately on data sets A and B. Then, for each index, average values were computed and consequent ranks were determined (lowest rank indicating best performance). In the last two columns, Table 4.6 presents average ranks computed across the three previously obtained ranks, and overall ranks determined by ranking the average ranks. It can be seen that the rankings obtained according to the three indices are almost identical with each other. According to external criterion validity, the best ranking methods were centroid linkage (6.A) and k-means methods using Euclidean distances (7.B) and city-block distances (7.C). Also performing very well and having values quite similar to the top ranked methods were complete linkage using correlation coefficients (5.A) and city-block distances (5.C), and ISODATA (8.X). At the other end of the spectrum of external criterion validity, the poorest performing methods were NORMAP/NORMIX (10.X) and Chernoff's faces (3.X) methods.

Table 4.6

Performance and rankings of quantitative taxonomic methods according to external criterion validity on the archetypal psychiatric patients data base

	Quantitative taxonomic method	% Concordance				Cramér's statistic				Correlation coefficient				Avg. rank	Overall rank
		Data set A	Data set B	Avg.	Rank	Data set A	Data set B	Avg.	Rank	Data set A	Data set B	Avg.	Rank		
1.A	Q-factor analysis, corr. coef.	93·18	93·18	93·18	10	0·915	0·918	0·916	11	0·815	0·826	0·821	11	10·66	10·5
2.A	Multidimen. scal., corr. coef.	77·28	97·93	87·61	12	0·862	0·972	0·917	10	0·710	0·938	0·824	10	10·66	10·6
2.B	Multidimen. scal., Euclid. dist.	85·23	96·59	90·91	11	0·852	0·960	0·905	12	0·697	0·910	0·804	12	11·66	12
2.C	Multidimen. scal., city-bl. dist.	75·00	100·00	87·50	13	0·788	1·000	0·888	13	0·592	1·000	0·796	13	13·00	13
3.X	Chernoff's faces	62·50	59·09	60·80	17	0·655	0·665	0·659	17	0·435	0·424	0·430	17	17·00	17
4.A	Single linkage, corr. coef.	52·27	72·73	62·50	15	0·605	0·816	0·710	16	0·519	0·698	0·609	14	15·00	15
4.B	Single linkage, Euclid. dist.	52·57	77·27	64·77	14	0·616	0·826	0·721	14	0·497	0·717	0·607	15	14·33	14
4.C	Single linkage, city-bl. dist.	52·27	70·54	61·36	16	0·616	0·816	0·716	15	0·497	0·670	0·584	16	15·66	16
5.A	Complete linkage, corr. coef.	97·73	97·73	97·73	5	0·972	0·972	0·972	5	0·938	0·938	0·938	5	5·00	5
5.B	Complete linkage, Euclid. dist.	90·91	97·73	94·32	9	0·907	0·972	0·939	9	0·793	0·938	0·866	9	9·00	9
5.C	Complete linkage, city-bl. dist.	97·73	97·73	97·73	5	0·972	0·972	0·972	5	0·938	0·938	0·938	5	5·00	5
6.A	Centroid linkage, corr. coef.	100·00	97·73	98·87	2	1·000	0·972	0·986	2	1·000	0·938	0·969	2	2·00	2
7.A	*k*-Means, corr. coef.	93·18	97·73	95·46	8	0·917	0·972	0·944	8	0·821	0·938	0·880	8	8·00	8
7.B	*k*-Means, Euclid. dist.	97·73	100·00	98·87	2	0·972	1·000	0·986	2	0·938	1·000	0·969	2	2·00	2
7.C	*k*-Means, city-bl. dist.	97·73	100·00	98·87	2	0·972	1·000	0·986	2	0·938	1·000	0·969	2	2·00	2
8.X	ISODATA	97·73	97·73	97·73	5	0·972	0·972	0·972	5	0·938	0·938	0·938	5	5·00	5
9.X	Rubin-Friedman	95·45	97·73	96·59	7	0·943	0·972	0·958	7	0·876	0·938	0·907	7	7·00	7
10.X	NORMAP/NORMIX	40·91	47·73	44·32	18	0·364	0·336	0·350	18	0·072	0·043	0·058	18	18·00	18

Next to the poorest performing methods were the single linkage methods (4.A, 4.B and 4.C).

Table 4.7 presents the performance values and rankings of the quantitative taxonomic methods according to internal criterion validity on the archetypal psychiatric patients data base. The cophenetic correlation values for each taxonomic method on data sets A and B and their average as well as overall ranks are given. According to internal criterion validity the best ranked methods turn out to be single linkage (4.A, 4.B and 4.C) and ordinal multidimensional scaling using correlation coefficients (2.A). Also well ranked, with absolute cophenetic correlation coefficients quite similar to the top ones, were *k*-means methods using Euclidean distances (7.B) and correlation coefficients (7.A), complete linkage methods using Euclidean distances (5.B) and correlation coefficients (5.A), and ISODATA (8.X). The poorest ranked methods were NORMAP/NORMIX (10.X) and Chernoff's faces (3.X).

Table 4.7

Performance and rankings of quantitative taxonomic methods according to internal criterion validity on the archetypal psychiatric patients data base

Quantitative taxonomic method		Cophenetic correlation			Rank
		Data set A	Data set B	Average	
1.A	Q-factor analysis, corr. coef.	0·623	0·623	0·6230	16
2.A	Multidimen. scal., corr. coef.	0·743	0·738	0·7405	3
2.B	Multidimen. scal., Euclid. dist.	0·683	0·721	0·7025	13
2.C	Multidimen. scal., city-bl. dist.	0·669	0·715	0·6920	14
3.X	Chernoff's faces	0·639	0·584	0·6112	17
4.A	Single linkage, corr. coef.	0·824	0·844	0·8340	1
4.B	Single linkage, Euclid. dist.	0·738	0·759	0·7445	2
4.C	Single linkage, city-bl. dist.	0·742	0·715	0·7285	4
5.A	Complete linkage, corr. coef.	0·695	0·733	0·7140	8·5
5.B	Complete linkage, Euclid. dist.	0·693	0·738	0·7155	6·5
5.C	Complete linkage, city-bl. dist.	0·696	0·712	0·7040	12
6.A	Centroid linkage, corr. coef.	0·695	0·675	0·6850	15
7.A	*k*-Means, corr. coeff.	0·698	0·733	0·7155	6·5
7.B	*k*-Means, Euclid. dist.	0·717	0·737	0·7270	5
7.C	*k*-Means, city-bl. dist.	0·696	0·715	0·7055	11
8.X	ISODATA	0·695	0·733	0·7140	8·5
9.X	Rubin-Friedman	0·690	0·733	0·7115	10
10.X	NORMAP/NORMIX	0·073	0·032	0·0525	18

Table 4.8 shows the replicability performance and ranking of the quantitative taxonomic methods on the archetypal psychiatric patients data base. The correlation coefficient for each taxonomic method, computed between corresponding entries of the two expert-method cross-classification

Table 4.8
Performance and rankings of quantitative taxonomic methods according to replicability on the archetypal psychiatric patients data base

Quantitative taxonomic method	Correlation coefficient (between expert-method cross-classification tables)	Rank
1.A Q-factor analysis, corr. coef.	0·971	10
2.A Multidimen. scal., corr. coef.	0·782	13
2.B Multidimen. scal., Euclid. dist.	0·891	11
2.C Multidimen. scal., city-bl. dist.	0·780	14
3.X Chernoff's faces	0·438	17
4.A Single linkage, corr. coef.	0·758	15
4.B Single linkage, Euclid. dist.	0·068	18
4.C Single linkage, city-bl. dist.	0·786	12
5.A Complete linkage, corr. coef.	1·000	2
5.B Complete linkage, Euclid. dist.	0·974	9
5.C Complete linkage, city-bl. dist.	1·000	2
6.A Centroid linkage, corr. coef.	0·998	5
7.A *k*-Means, corr. coef.	0·995	8
7.B *k*-Means, Euclid. dist.	0·998	5
7.C *k*-Means, city-bl. dist.	0·998	5
8.X ISODATA	1·000	2
9.X Rubin-Friedman	0·997	7
10.X NORMAP/NORMIX	0·441	16

tables from data sets A and B, and the consequent ranks, are presented. According to clustering replicability across the randomly produced halves of the original data base, the best ranked methods were complete linkage methods using correlation coefficients (5.A) and city-block distances (5.C), and ISODATA (8.X). Also very well ranked, with absolute replicability coefficients of 0·998, were centroid linkage (6.A) and *k*-means methods using Euclidean distances (7.B) and city-block distances (7.C). The poorest ranked methods were single linkage using Euclidean distances (4.B), Chernoff's faces (3.X) and NORMAP/NORMIX (10.X).

The rankings of quantitative taxonomic methods according to replicability and external criterion validity (our two cleanest evaluative measures) were quite similar to each other. In fact, the Spearman rank correlation coefficient computed between them is 0·90.

Table 4.9 presents the ranking of the quantitative taxonomic methods across all three major evaluative criteria (external criterion validity, internal criterion validity and replicability) on the archetypal psychiatric patients data base. The best ranked methods turned out to be *k*-means methods using Euclidean distances (7.B) and city-block distances (7.C), complete

Table 4.9

Ranking of the quantitative taxonomic methods across all three major evaluative criteria (external criterion validity, internal criterion validity and replicability) on the archetypal psychiatric patients data base

Quantitative taxonomic method	ECV rank	ICV rank	Replic. rank	Average rank	Overall rank
1.A Q-factor analysis, corr. coef.	10·5	16	10	12·17	15
2.A Multidimen. scal., corr. coef.	10·5	3	13	8·83	10
2.B Multidimen. scal., Euclid. dist.	12	13	11	12·00	14
2.C Multidimen. scal., city-bl. dist.	13	14	14	13·67	16
3.X Chernoff's faces	17	17	17	17·00	17
4.A Single linkage, corr. coef.	15	1	15	10·33	11
4.B Single linkage, Euclid. dist.	14	2	18	11·33	13
4.C Single linkage, city-bl. dist.	16	4	12	10·67	12
5.A Complete linkage, corr. coef.	5	8·5	2	5·17	2·5
5.B Complete linkage, Euclid. dist.	9	6·5	9	8·17	9
5.C Complete linkage, city-bl. dist.	5	12	2	6·33	5
6.A Centroid linkage, corr. coef.	2	15	5	7·33	6
7.A *k*-Means, corr. coef.	8	6·5	8	7·50	7
7.B *k*-Means, Eculid. dist.	2	5	5	4·00	1
7.C *k*-Means, city-bl. dist.	2	11	5	6·00	4
8.X ISODATA	5	8·5	2	5·17	2·5
9.X Rubin-Friedman	7	10	7	8·00	8
10.X NORMAP/NORMIX	18	18	16	17·33	18

linkage using correlation coefficients (5.A) and city-block distances (5.C), ISODATA (8.X) and centroid linkage (6.A). Clearly, the poorest ranked methods were NORMAP/NORMIX (10.X) and Chernoff's faces (3.X).

Table 4.10 shows the performance and ranking of the ordinal multidimensional scaling and Chernoff's faces methods according to inter-rater agreement on the archetypal psychiatric patients data base. For each one of these taxonomic methods, the concordance percentage, Cramér's statistic and correlation coefficient values for data sets A and B and their average ranks computed across the previously obtained ranks, and overall ranks are presented. It can be seen that the rankings obtained according to these three inter-rater agreement indices are quite similar to each other. The three ordinal multidimensional scaling methods (2.A, 2.B and 2.C) have similar absolute inter-rater agreement values with each other and ranked better, in the listed order, than Chernoff's faces (3.X).

In general agreement with aspects of this study, Strauss, Bartko and Carpenter (1973) found that the complete linkage method produced clusterings of archetypal psychiatric patients closer to a classification established by experienced psychiatrists, than did the Rubin-Friedman optimization technique.

Table 4.10

Performance and ranking of the ordinal multidimensional scaling and Chernoff's faces methods according to inter-rater agreement on the archetypal psychiatric patients data base

	Quantitative taxonomic method	% Concordance				Cramér's statistic				Correlation coefficient				Avg. rank	Overall rank
		Data set A	Data set B	Avg.	Rank	Data set A	Data set B	Avg.	Rank	Data set A	Data set B	Avg.	Rank		
2.A	Multidimen. scal., Corr. Coef.	72·73	100·00	86·37	1	0·816	1·000	0·908	1	0·668	1·000	0·834	1	1	1
2.B	Multidimen. scal., Euclid. dist.	75·00	93·18	84·09	2	0·837	0·927	0·882	2	0·677	0·833	0·755	3	2·33	2
2.C	Multidimen. scal., city-bl. dist.	77·27	80·00	78·64	3	0·760	0·746	0·753	3	0·610	1·000	0·805	2	2·66	3
3.X	Chernoff's faces	56·82	59·09	57·96	4	0·554	0·489	0·521	4	0·383	0·295	0·339	4	4	4

In the light of the consistent ranking patterns noted in this study, it is reasonable to recommend nearest centroid sorting (k-means and ISODATA) and complete linkage as the preferred methods for clustering psychiatric data. If these procedures are not available to the user, the centroid linkage method and the Rubin-Friedman optimization technique seem to be appropriate alternatives.

The use of either multivariate normal mixture analysis (NORMAP/NORMIX) or facial representation of multidimensional points (Chernoff's faces) for clustering of psychiatric data, without previous successful testing on data of the specific type to be analyzed, does not seem to be warranted.

It should be noted that the results of a quantitative taxonomic study may be influenced not only by the cluster analytic method used (as shown in this study), but also by the scaling and transformation of the input data (as shown by Bartko, Strauss and Carpenter, 1971). Illustrative of this concern for stability is the recent study of schizophrenic subtypes conducted by Carpenter *et al.* (1976), in which the authors were able to ascertain the presence of four major subtypes after using two different clustering methods (complete linkage and k-means), two different starting configurations for the k-means method, and seven instances of dropping three randomly selected patients from their data set of 16 composite patients.

As indicated earlier, the results of the present study suggest preferred methods for conducting taxonomic studies on psychiatric patients in general. If a narrower problem (e.g. typology of affective disorders) is under consideration, the analysis of a set of pertinent archetypal patients for more specifically selecting a clustering approach seems to be advisable.

4.5 PERFORMANCE OF RELATIONSHIP MEASURES

Table 4.11 exhibits the performance and ranking of the relationship measures (correlation coefficient, Euclidean distance and city-block distance), according to external criterion validity, internal criterion validity and replicability, and across these three criteria. Each evaluative criterion value was computed by averaging corresponding values of the quantitative taxonomic methods (2, 4, 5 and 7) which used all three relationship measures. The three relationship measures were ranked according to each of the three external criterion validity indices, and these ranks were averaged to produce overall average ranks according to this evaluative criterion. These and the internal criterion validity and replicability ranks were then averaged, and the results are shown in the last two columns of Table 4.11.

The ranking was not conclusive, as the magnitudes of the relationship measures on each evaluative criterion were either identical or quite similar

Table 4.11

Performance and ranking of relationship measures according to external criterion validity (ECV), internal criterion validity (ICV) and replicability on the archetypal psychiatric patients data base

Relationship measure	ECV				ICV		Replic.		Average rank	Overall rank
	% Concordance	Cramér	Corr. coef.	Overall average rank	Cophenetic rank corr.	Rank	Corr. coef.	Rank		
A Correlation coefficient	85·83	0·89	0·81	3	0·75	1	0·88	2	2·00	2
B Euclidean distance	87·22	0·89	0·81	1·5	0·72	2	0·73	3	2·17	3
C City-block distance	86·37	0·89	0·82	1·5	0·71	3	0·89	1	1·83	1

to each other. This was in line with the quite varied performance of the relationship measures when used with particular clustering techniques. Consequently, no strong recommendation can be given regarding preferred relationship measures in psychiatric diagnostic data.

4.6 CLUSTER-BY-CLUSTER AGREEMENT BETWEEN EXPERT AND TAXONOMIC METHODS

The quantitative taxonomic methods used in this study for grouping archetypal or conceptual patients represent quite different approaches to the classification problem. Their variety may resemble the variety observed among human judges with different theoretical backgrounds and cognitive styles, undertaking the task of developing a 'clinically useful' classification of psychiatric patients. On these grounds, the study of the agreement between groupings produced by several quantitative taxonomic methods and those generated by clinicians, can be seen as a kind of reliability study. This is also in line with the relationship generally noted between the degree of structural overlapping of two diagnostic categories and their lack of inter-rater reliability (Blashfield and Draguns, 1976).

This approach to the study of the structural clarity of diagnostic categories was implemented by assessing the cluster-by-cluster agreement between expert and taxonomic methods. The classification produced by each clustering method was cross-tabulated with the clinicians' classification underlying the archetypal patient data base. The resulting 44 cross-classification tables (22 for each data set, including two tables for each of the three ordinal multidimensional scaling methods and the Chernoff's faces method as two judges produced separate clusterings for each of these methods) were summed cell-by-cell. Table 4.12 presents these summary cross-classification figures.

From Table 4.12, the percentages of each one of the clinician-developed groups classified into the various clusters produced by the taxonomic methods were computed and are presented in Table 4.13.

The highest level of agreement between clinicians and clustering procedures was found for manic-depressive depression (92·7% of the corresponding archetypal patients); the second for paranoid schizophrenia (88·4%); the third for simple schizophrenia (81·6%); and the lowest for mania (75·6%). Two forms of misclassification accounted for most of the clinician-computer disagreement. One was the confusion between mania and paranoid schizophrenia (22·3% of the manics and 9·5% of the paranoid schizophrenics experienced this form of misclassification). The other was the confusion between depression and simple schizophrenia (4·1% of the

Table 4.12
Cell-by-cell sum of cross-classification tables between clinicians and quantitative taxonomic methods on the archetypal psychiatric patients data base

Clinician-produced groups	Clusters produced by quantitative taxonomic methods: Manic-depressive depressed	Manic-depressive manic	Simple schizophrenic	Paranoid schizophrenic	
Manic-depressive depressed	444	12	20	8	484
Manic-depressive manic	4	366	6	108	484
Simple schizophrenic	65	11	395	13	484
Paranoid schizophrenic	5	46	5	428	484
	518	435	426	557	

depressives and 13·4% of the simple schizophrenics were subject to this confusion).

One of the most conspicuous results of this study was the considerable degree of misclassification or overlapping observed between the manic and the paranoid schizophrenic groups of archetypal patients. This was

Table 4.13
Percentages of each one of the clinician-produced groups classified into the various clusters produced by the quantitative taxonomic methods on the archetypal psychiatric patients data base

Clinician-produced groups	Clusters produced by quantitative taxonomic methods: Manic-depressive depressed	Manic-depressive manic	Simple schizophrenic	Paranoid schizophrenic	
Manic-depressive depressed	92·74	2·78	4·13	1·65	100·00
Manic-depressive manic	0·82	75·62	1·24	22·31	100·00
Simple schizophrenic	13·43	2·27	81·61	2·69	100·00
Paranoid schizophrenic	1·03	9·50	1·03	88·43	100·00

consistent with the geometrical proximity noted between these groups on the ordinal multidimensional scaling plots (Figs 4.2 and 4.3) and the clustering of one manic archetypal patient with the 11 paranoid schizophrenics obtained through complete linkage (Figs 4.4 and 4.5), ISODATA (Table 4.4) and the Rubin-Friedman reallocation method (Table 4.5). Such proximity and overlapping is well acknowledged by clinical experience. Testimonies of this are both the inclusion of a schizoaffective disorder category in the current American classification of mental disorders (DSM-III, American Psychiatric Association, 1980), and systematic reports in the literature about patients admitted to hospital with the diagnosis of paranoid schizophrenia who upon thorough examination are rediagnosed as manics (Abrams, Taylor and Gastanaga, 1974).

5

Cluster Analysis of Iris Specimens

5.1 INTRODUCTION

The data bases studied in the preceding two chapters were composed of treatment environments, as perceived by their inhabitants, and archetypal psychiatric patients defined on symptoms and signs. The respective psychosocial and psychopathological contents of these data are good examples of behavioral or psychological information, both in terms of measurement and classification. They reflect the currently developing status of the behavioral sciences: incipient, tentative, experimental. The data studied in this chapter, iris specimens measured on morphological variables, are somewhat 'harder'. They correspond to the long tradition of biological, more specific cally botanical, taxonomy. Early investigators in this field, such as Carolus Linnaeus, have also been influential in inspiring and promoting the search for order and structure in other fields.

Although the purposes of taxonomy in behavioral science are somewhat different from those in botany, there are some commonalities between these fields. Commonalities go beyond the generality of the methodology used for creating or discovering groups, to include factors underlying the established or proposed classifications. For example, on the one hand it has been reported in the literature, and will be suggested further in this chapter, that genetic factors have a great deal to do with the grouping of iris specimens, and that the grouping outcome, in turn, can help to generate and elucidate hypotheses about the genetic structure of the studied specimens. On the other hand, many behavioral scientists are significantly interested in studying the role of genetic factors in explaining (probably in a partial and predispositional way) human behavior, both normal and abnormal. For example, there is increasing information which suggests that genetic factors have a significant although limited role in the development of some psychiatric disorders such as schizophrenia (Rosenthal and Ketty, 1968) and manic-depressive illness (Winokur, Clayton and Reich, 1969). It is also likely that the development of better groupings of psychiatric patients will facilitate the elucidation of etiological factors, including genetic and constitutional ones.

A data base composed of 150 iris specimens belonging to three species (*Iris setosa, Iris versicolor* and *Iris virginica*) and measured on sepal length and width and petal length and width, was used in this cluster analytic study. Groupings of these specimens were produced by using the quantitative taxonomic methods described in § 2.2. Examples of obtained clustering processes and outcomes are presented for illustrative purposes. The comparative performance of the quantitative taxonomic methods and relationships measures on this data base, evaluated according to criteria described in § 2.3, is then shown and discussed. Finally, the levels and areas of agreement and disagreement between the established botanical classification and the groupings of iris specimens developed through the various quantitative taxonomic methods, are presented.

5.2 DATA BASE

The iris data used in this study were published by Fisher (1936) in his well-known paper on discriminant functions. This data set has served a number of times since then for testing taxonomic methods proposed by other investigators (e.g. Friedman and Rubin, 1967; Kendall, 1966; Solomon, 1971).

The data base is composed of 150 iris specimens each measured on four morphological variables: sepal length and width and petal length and width. The species *I. setosa, I. versicolor* and *I. virginica* were represented by 50 plants each. The first two groups were found growing together in the same colony and were measured by the botanist E. Anderson, while the *I. virginica* specimens were taken from a different colony. Randolph (1934) accertained, and Anderson confirmed, that *I. setosa* is a diploid species with 38 chromosomes, *I. versicolor* is tetraploid with 70, and *I. versicolor* is hexaploid. Randolph suggested the possibility that *I. versicolor*, which is intermediate in three measurements although not in sepal width, is a polyploid hybrid of the two other species. Fisher (1936) found through discriminant function analysis that *I. setosa* could be clearly separated from the other two species, which overlapped with each other to some extent. Kendall (1966), using a clustering method utilizing only the rank order of the measurements, was able to recover all *I. setosa* specimens, but had difficulties in classifying a dozen plants from the *I. versicolor* and *I. virginica* groups. Friedman and Rubin (1967), using their hill-climbing procedure which maximizes various variance-covariance criteria, were always able to recover *I. setosa* well. When using the minimum trace **W** criterion, ten plants corresponding to *I. versicolor* and *I. virginica* were misclassified between these two groups while, when using both the maximum $|\mathbf{T}|/|\mathbf{W}|$ and the trace $\mathbf{W}^{-1}\mathbf{B}$ criteria, three plants from these two groups were confused in the same way. Solomon (1971), applying King's (1966) centroid linkage method of the iris data base, clustered together 48 of the 50 *I. setosa* specimens but found considerable overlap between the two other species.

As the replicability of cluster configurations produced by the various quantitative taxonomic methods was one of our evaluative criteria, two comparably composed data sets were required. These two data sets, labeled A and B, were obtained by randomly halving each one of the 50-specimen groups representing the three iris species.

Tables 5.1 and 5.2 present data matrices constituted by the scores of each of the 75 iris specimens of data sets A and B, respectively, on each of the four morphological variables.

Table 5.3 presents, separately for data sets A and B, the means and standard deviations of the three iris groups on the four morphological variables. Figure 5.1 shows the means profiles of the three iris groups from

Table 5.1
Data matrix of iris specimens measured on four morphological dimensions: data set A

Iris specimens		Sepal length	Sepal width	Petal length	Petal width
I. setosa	1	5·1	3·5	1·4	0·3
	2	4·4	3·2	1·3	0·2
	3	4·4	3·0	1·3	0·2
	4	5·0	3·5	1·6	0·6
	5	5·1	3·8	1·6	0·2
	6	4·9	3·1	1·5	0·2
	7	5·0	3·2	1·2	0·2
	8	4·6	3·2	1·4	0·2
	9	5·0	3·3	1·4	0·2
	10	4·8	3·4	1·9	0·2
	11	4·8	3·0	1·4	0·1
	12	5·0	3·5	1·3	0·3
	13	5·1	3·3	1·7	0·5
	14	5·0	3·4	1·5	0·2
	15	5·1	3·8	1·9	0·4
	16	4·9	3·0	1·4	0·2
	17	5·3	3·7	1·5	0·2
	18	4·3	3·0	1·1	0·1

Table 5.1 cont.

Iris specimens		Sepal length	Sepal width	Petal length	Petal width
	19	5·5	3·5	1·3	0·2
	20	4·8	3·4	1·6	0·2
	21	5·2	3·4	1·4	0·2
	22	4·8	3·1	1·6	0·2
	23	4·9	3·6	1·4	0·1
	24	4·6	3·1	1·5	0·2
	25	5·7	4·4	1·5	0·4
I. versicolor	26	6·4	3·2	4·5	1·5
	27	5·5	2·4	3·8	1·1
	28	5·7	2·9	4·2	1·3
	29	5·7	3·0	4·2	1·2
	30	5·6	2·9	3·6	1·3
	31	7·0	3·2	4·7	1·4
	32	6·8	2·8	4·8	1·4
	33	6·1	2·8	4·7	1·2
	34	4·9	2·4	3·3	1·0
	35	5·8	2·7	3·9	1·2
	36	5·8	2·6	4·0	1·2
	37	5·5	2·4	3·7	1·0
	38	6·7	3·0	5·0	1·7
	39	5·7	2·8	4·1	1·3
	40	6·7	3·1	4·4	1·4
	41	5·5	2·3	4·0	1·3
	42	5·1	2·5	3·0	1·1
	43	6·6	2·9	4·6	1·3
	44	5·0	2·3	3·3	1·0
	45	6·9	3·1	4·9	1·5
	46	5·0	2·0	3·5	1·0
	47	5·6	3·0	4·5	1·5
	48	5·6	3·0	4·1	1·3
	49	5·8	2·7	4·1	1·0
	50	6·3	2·3	4·4	1·3
I. virginica	51	6·3	3·3	6·0	2·5
	52	6·7	3·3	5·7	2·1
	53	7·2	3·6	6·1	2·5
	54	7·7	3·8	6·7	2·2
	55	7·2	3·0	5·8	1·6
	56	7·4	2·8	6·1	1·9
	57	7·6	3·0	6·6	2·1
	58	7·7	2·8	6·7	2·0
	59	6·2	3·4	5·4	2·3
	60	7·7	3·0	6·1	2·3
	61	6·8	3·0	5·5	2·1
	62	6·4	2·7	5·3	1·9

Table 5.1 cont.

Iris specimens		Sepal length	Sepal width	Petal length	Petal width
I. virginica	63	5·7	2·5	5·0	2·0
	64	6·9	3·1	5·1	2·3
	65	5·9	3·0	5·1	1·8
	66	6·3	3·4	5·6	2·4
	67	5·8	2·7	5·1	1·9
	68	6·3	2·7	4·9	1·8
	69	6·0	3·0	4·8	1·8
	70	7·2	3·2	6·0	1·8
	71	6·2	2·8	4·8	1·8
	72	6·9	3·1	5·4	2·1
	73	6·7	3·1	5·6	2·4
	74	6·4	3·1	5·5	1·8
	75	5·8	2·7	5·1	1·9

Table 5.2
Data matrix of iris specimens measured on four morphological dimensions: data set B

Iris specimens		Sepal length	Sepal width	Petal length	Petal width
I. setosa	1	5·7	3·8	1·7	0·3
	2	4·8	3·0	1·4	0·3
	3	5·2	4·1	1·5	0·1
	4	4·7	3·2	1·6	0·2
	5	4·5	2·3	1·3	0·3
	6	5·4	3·4	1·7	0·2
	7	5·0	3·0	1·6	0·2
	8	4·6	3·4	1·4	0·3
	9	5·4	3·9	1·3	0·4
	10	5·0	3·6	1·4	0·2
	11	5·4	3·9	1·7	0·4
	12	4·6	3·6	1·0	0·2
	13	5·1	3·8	1·5	0·3
	14	5·8	4·0	1·2	0·2
	15	5·4	3·7	1·5	0·2
	16	5·0	3·4	1·6	0·4
	17	5·4	3·4	1·5	0·4
	18	5·1	3·7	1·5	0·4

Table 5.2 cont.

Iris specimens		Sepal length	Sepal width	Petal length	Petal width
	19	4·4	2·9	1·4	0·2
	20	5·5	4·2	1·4	0·2
	21	5·1	3·4	1·5	0·2
	22	4·7	3·2	1·3	0·2
	23	4·9	3·1	1·5	0·1
	24	5·2	3·5	1·5	0·2
	25	5·1	3·5	1·4	0·2
I. versicolor	26	6·1	3·0	4·6	1·4
	27	5·9	3·0	4·2	1·5
	28	6·0	2·7	5·1	1·6
	29	5·6	2·5	3·9	1·1
	30	6·7	3·1	4·7	1·5
	31	6·2	2·2	4·5	1·5
	32	5·9	3·2	4·8	1·8
	33	6·3	2·5	4·9	1·5
	34	6·0	2·9	4·5	1·5
	35	5·6	2·7	4·2	1·3
	36	6·2	2·9	4·3	1·3
	37	6·0	3·4	4·5	1·6
	38	6·5	2·8	4·6	1·5
	39	5·7	2·8	4·5	1·3
	40	6·1	2·9	4·7	1·4
	41	5·5	2·5	4·0	1·3
	42	5·5	2·6	4·4	1·2
	43	5·4	3·0	4·5	1·5
	44	6·3	3·3	4·7	1·6
	45	5·2	2·7	3·9	1·4
	46	6·4	2·9	4·3	1·3
	47	6·6	3·0	4·4	1·4
	48	5·7	2·6	3·5	1·0
	49	6·1	2·8	4·0	1·3
	50	6·0	2·2	4·0	1·0
I. virginica	51	6·1	3·0	4·9	1·8
	52	6·0	2·2	5·0	1·5
	53	6·4	3·2	5·3	2·3
	54	5·8	2·8	5·1	2·4
	55	6·9	3·2	5·7	2·3
	56	6·7	3·0	5·2	2·3
	57	7·7	2·6	6·9	2·3
	58	6·3	2·8	5·1	1·5
	59	6·5	3·0	5·2	2·0
	60	7·9	3·8	6·4	2·0
	61	6·1	2·6	5·6	1·4
	62	6·4	2·8	5·6	2·1

Table 5.2 cont.

Iris specimens		Sepal length	Sepal width	Petal length	Petal width
I. virginica	63	6·3	2·5	5·0	1·9
	64	4·9	2·5	4·5	1·7
	65	6·8	3·2	5·9	2·3
	66	7·1	3·0	5·9	2·1
	67	6·7	3·3	5·7	2·5
	68	6·3	2·9	5·6	1·8
	69	6·5	3·0	5·5	1·8
	70	6·5	3·0	5·8	2·2
	71	7·3	2·9	6·3	1·8
	72	6·7	2·5	5·8	1·8
	73	5·6	2·8	4·9	2·0
	74	6·4	2·8	5·6	2·2
	75	6·5	3·2	5·1	2·0

Table 5.3
Means and standard deviations of the three iris groups, on four measurements: data sets A and B

Iris group	Statistic	Sepal length	Sepal width	Petal length	Petal width
Data set A					
I. setosa	Mean	4·932	3·376	1·468	0·240
(N=25)	SD	0·324	0·324	0·189	0·119
I. cersicolor	Mean	5·892	2·732	4·132	1·260
(N=25)	SD	0·626	0·329	0·533	0·187
I. virginica	Mean	6·680	3·044	5·600	2·052
(N=25)	SD	0·647	0·307	0·572	0·254
Data set B					
I. setosa	Mean	5·080	3·480	1·456	0·252
(N=25)	SD	0·371	0·427	0·161	0·092
I. versicolor	Mean	5·980	2·808	4·388	1·392
(N=25)	SD	0·385	0·300	0·364	0·189
I. virginica	Mean	6·496	2·904	5·504	2·000
(N=25)	SD	0·624	0·328	0·538	0·297

data sets A and B. It can be seen that in both data sets the mean profile of *I. versicolor* looks parallel to that of *I. virginica* and closer to it than to that of *I. setosa*. Additionally, each of the mean scores of *I. versicolor*, except that for sepal width, are intermediate between those of *I. virginica* and *I. setosa*.

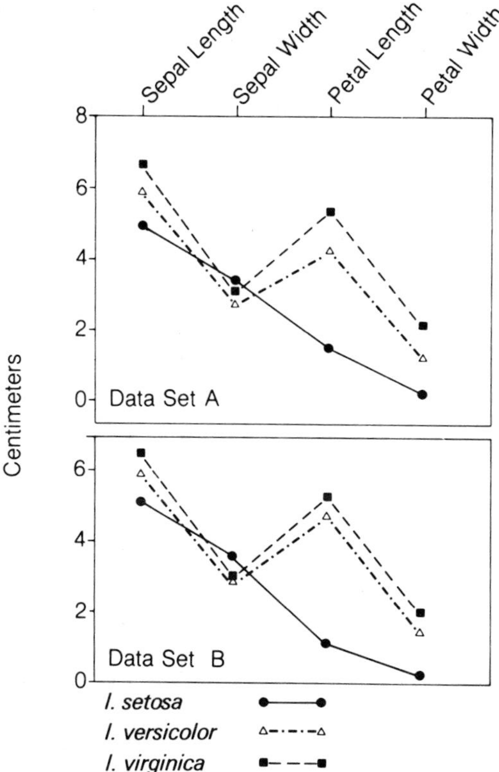

Fig. 5.1 Profiles of three iris groups of data sets A and B.

5.3 ILLUSTRATIVE CLUSTER ANALYTIC RESULTS

The quantitative taxonomic methods listed in Table 2.1 were applied to data sets A and B of iris specimens. A group of figures and a table will be presented in this section, containing clustering processes and outcomes obtained from the application of some of the taxonomic methods on this data base. The comparative rankings of the methods will be described in § § 5.4.

As was done for the treatment environment and archetypal psychiatric patient data bases, and as will be done in Chapter 6 for ethnic populations, the results from the application of complete linkage hierarchical cluster analysis and ordinal multidimensional scaling, both using correlation coefficients, on data sets A and B will be presented first. Then, results from applying Chernoff's faces and pattern clustering by multivariate mixture analysis on data sets A and B, respectively, will be described. The intention of presenting two different taxonomic methods, in addition to complete linkage and ordinal multidimensional scaling for each data base is to illustrate the use of the ten basic clustering approaches.

Figures 5.2 and 5.3 show two-dimensional representations of the iris specimens of data sets A and B, respectively, obtained by using ordinal multidimensional scaling with correlation coefficients. Quite low stress values (0·012 and 0·016) were obtained for both bidimensional configurations, indicating close relationships between the rankings of interpoint distances in the resulting configuration and the input data. Each specimen is represented by a symbol indicating the species to which it belongs. In both data sets, the 25 specimens of *I. setosa* form a clear cluster, quite separated from a conglomerate of the other 50 specimens, so that if the specimens were unlabeled two major clusters would be visualized. A closer look at the larger conglomerate of points in both data sets shows that most of the *I. versicolor* specimens, on the one hand, and most of the *I. virginica* specimens, on the other hand, are clustered together as such, respectively forming two clumps adjacent to each other. In data set A, if one intermediate *I. versicolor* is omitted from consideration, it is possible to draw a vertical line separating the adjacent clumps into right and left. In data set B, there is an increasing mixture of *I. versicolor* and *I. virginica*, but it is still possible to visualize two adjacent clusters or clumps corresponding to these two species. It can also be seen that the *I. versicolor* clump forms the pole or half of the larger conglomerate oriented towards the distant *I. setosa* cluster, and therefore it is located in a position geometrically intermediate between the other two specimen groups. Two people were asked to inspect visually the plots presented in Figs 5.2 and 5.3 (which included at that time plain points unidentified according to species), and to draw boundaries dividing the 75 specimens into three groups. For data set A, both judges circled the 25 *I. setosa* into one cluster, one judge circled two and the other judge 18 *I. versicolor* specimens to form intermediately located clusters and, finally, both judges defined larger mixed clusters composed of the 25 *I. virginica* and the remainder of the *I. versicolor* specimens.

Figures 5.4 and 5.5 exhibit dendrograms representing complete linkage cluster anlyses, using correlation coefficients, of the iris specimens of data sets A and B. The entire clustering process is depicted for each data set, from the beginning where each specimen constitutes a singular cluster to the end where all specimens are lumped into one universal cluster. Figure 5.4 shows that, for data set A, the 25 *I. setosa* specimens form a cluster quite early in the process, at similarity level 0·735. At the same early level another

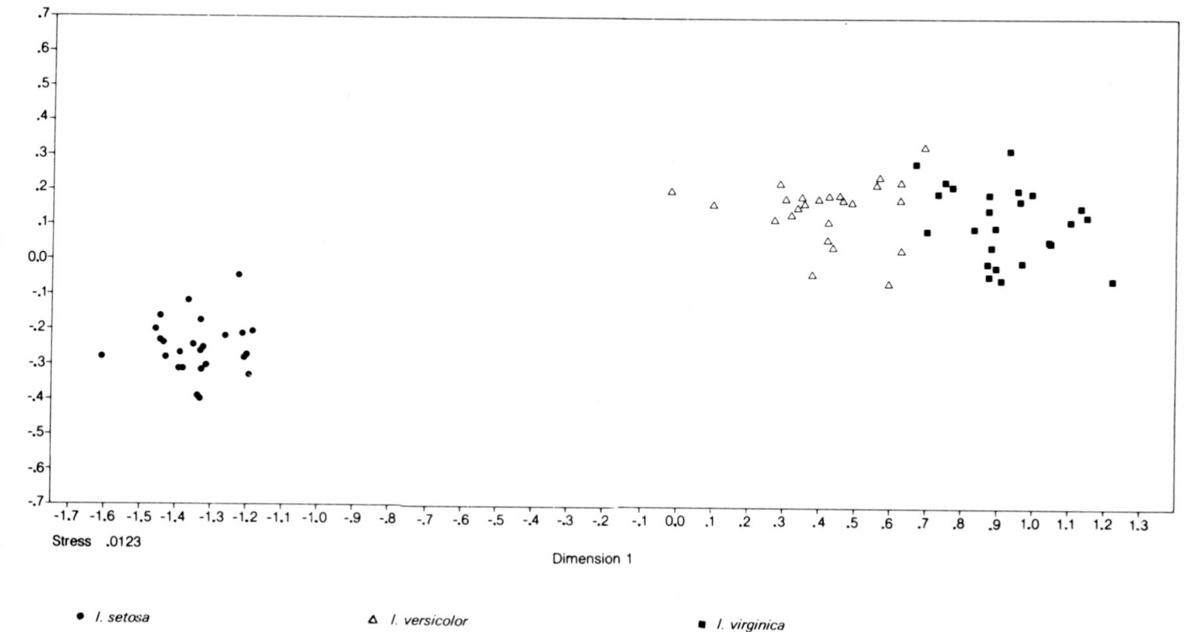

Fig. 5.2 Ordinal multidimensional scaling representation, using correlation coefficients, of the 75 iris specimens of data set A.

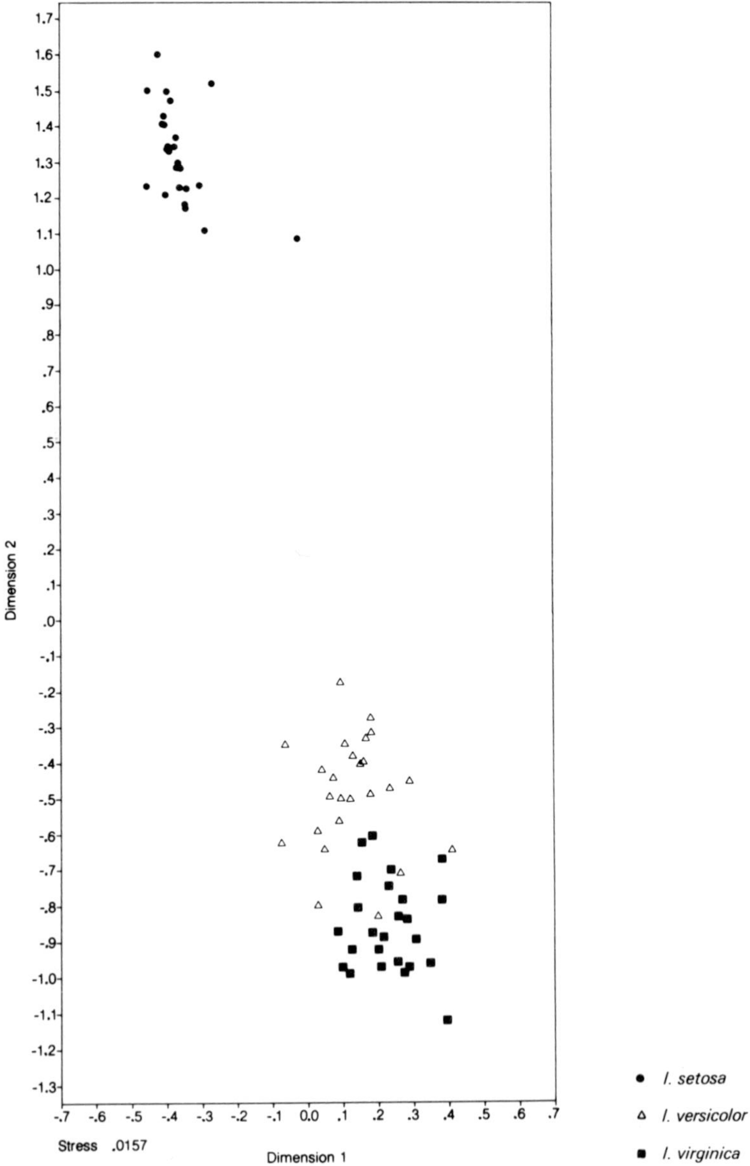

Fig. 5.3 Ordinal multidimensional scaling representation, using correlation coefficients of the 75 iris specimens of data set B.

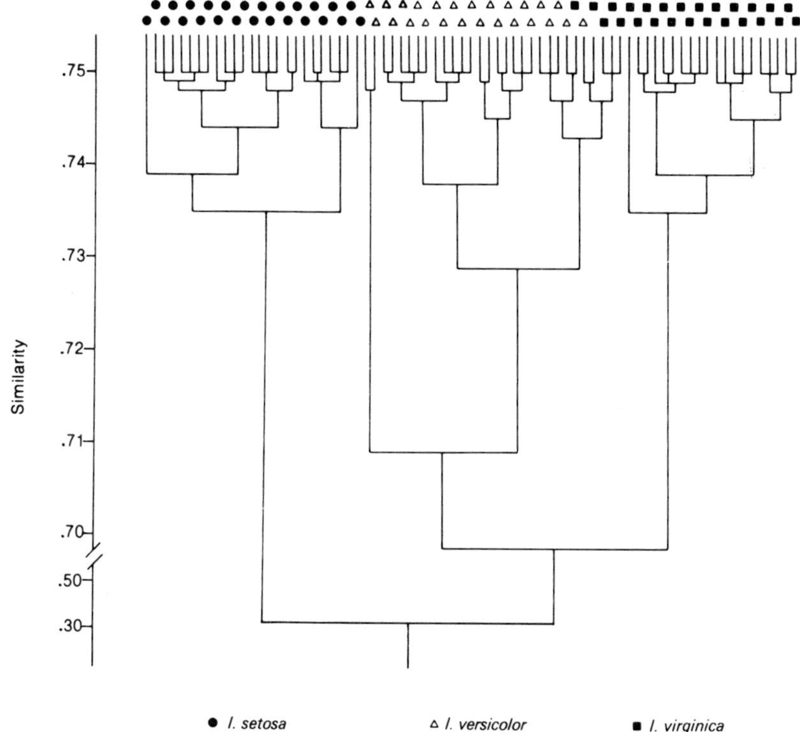

Fig. 5.4 Dendrogram representing a complete linkage cluster analysis, using correlation coefficients, of the 75 iris specimens of data set A.

persistent cluster is formed, constituted by 20 *I. virginica* specimens. Even earlier, at similarity level 0·743, a very transitory cluster developed, including the other five *I. virginica* plus five *I. versicolor* specimens. This cluster later joined other clusters exclusively composed of *I. versicolor* specimens to form, at level 0·709, a 30-member mixed but predominantly *I. versicolor* cluster. At the next step (similarity level 0·647), this cluster and that composed of 20 *I. virginica* specimens joined. This large and heterogeneous cluster joined the pure *I. setosa* cluster at a much lower similarity level (0·300), which indicates the considerable taxonomic separation between them. Figure 5.5 shows a clustering process for data set B grossly similar to that obtained for data set A. At similarity level 0·725, a major cluster composed of 22 *I. versicolor* and one *I. virginica* specimens developed and at level 0·722 another major cluster, constituted by 24 *I. virginica* and three *I. versicolor*, also developed. At similarity level 0·701,

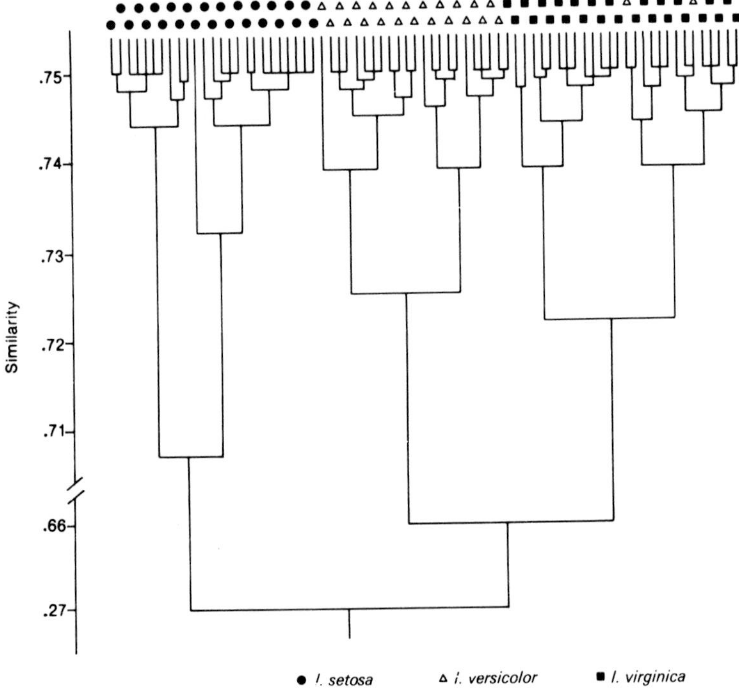

Fig. 5.5 Dendrogram representing a complete linkage cluster analysis, using correlation coefficients, of the 75 iris specimens of data set B.

the 25 *I. setosa* specimens agglomerated into one cluster. At the next step (similarity level 0·664), the first two clusters joined into a mixed *I. versicolor-I. virginica* cluster which, at a much lower level (0·268), was lumped into one cluster with all the *I. setosa* specimens.

Figure 5.6 shows Chernoff's faces representing the 75 iris specimens of data set A, arranged in groups according to their species identification. The relationship set between iris variables and face features was as follows: sepal length affected the distance from the nose center to the junction of the upper and lower ellipses of the face, sepal width affected the eccentricity of the upper ellipse of the face, petal length affected the position of the center of the mouth and petal width affected the curvature of the mouth. Two people were asked to group the 75 unidentified faces, each presented on a separate piece of paper, into three piles. For data set A, both judges put the 25 *I. setosa* plus some *i. versicolor* specimens (six the first judge and seven the second) into the same cluster. Also, both judges put the 25 *I. virginica*

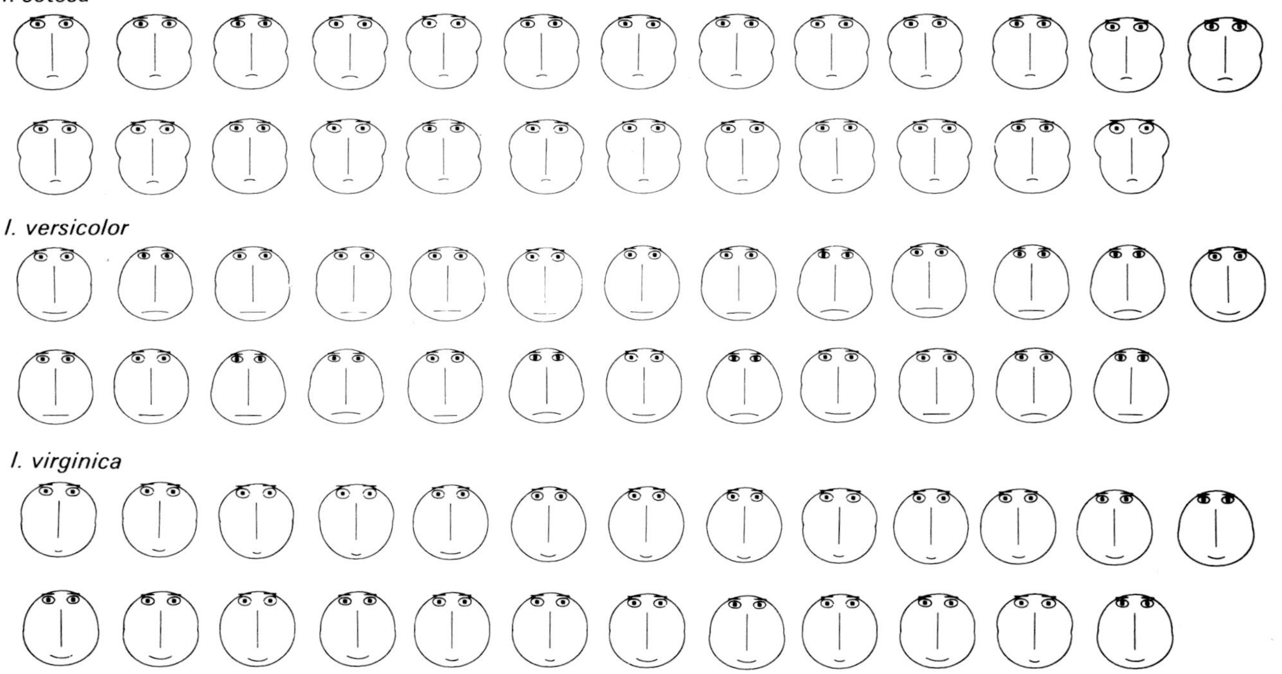

Fig. 5.6 Chernoff's faces representing the 75 iris specimens of data set A.

plus four *I. versicolor* specimens into one cluster. The remaining *I. versicolor* specimens were piled together by both judges. For data set B, both judges put together the 25 *I. setosa* plus three *I. versicolor* specimens into one cluster. They also put together most of the *I. versicolor* plus a few *I. virginica* specimens (20 and 3, respectively, by the first judge, and 19 and 1, respectively, by the second). The third cluster formed by both judges was composed predominantly of *I. virginica* plus a few *I. versicolor* specimens.

Table 5.4 outlines the results of applying pattern clustering by multivariate mixture analysis (NORMIX) to the 75 iris specimens of data set B. A comparison of the parameters of the three types at the last (twenty-ninth) iteration with the actual means and standard deviations of the three iris species in data set B presented in Table 5.3, shows that they are very similar for *I. setosa* but not so similar for the other two species. The final NORMIX solution put together the 25 *I. setosa* specimens under type I; 11 *I. versicolor* and 6 *I. virginica* specimens under type II; and 19 *I. virginica* and 14 *I. versicolor* under type III.

Table 5.4

Pattern clustering by multivariate mixture analysis (NORMIX) of 75 iris specimens of data set B

A Process started with *k*-means sorting for producing a starting 3-type partition

B Then, 29 iterations involving recalculations of the probabilities of membership of every point in each type were carried out until convergence

C At the twenty-ninth iteration the following parameters for the three types were obtained:

Proportion of population	Iris type	Statistic	Sepal length	Sepal width	Petal length	Petal width
0·33	Type I	Mean	5·080	3·481	1·456	0·252
		SD	0·363	0·418	0·154	0·090
0·23	Type II	Mean	6·461	2·715	4·901	1·486
		SD	0·624	0·379	0·944	0·330
0·44	Type III	Mean	6·110	2·930	4·968	1·806
		SD	0·501	0·238	0·565	0·374

The computation of probabilities of membership yielded the following 'most likely' partition:

Type I: 1-25

Type II: 29-31, 33, 36, 38, 46-50, 52, 57, 60, 61, 71, 72

Type III: 26-28, 34, 35, 37, 39-45, 51, 53-56, 58, 59, 62-70, 73-75

5.4 PERFORMANCE OF QUANTITATIVE TAXONOMIC METHODS

The comparative performance of the various quantitative taxonomic methods on the iris data base was evaluated according to external criterion validity, internal criterion validity, replicability and, for ordinal multidimensional scaling and Chernoff's faces, interjudge agreement, as well as across the first three evaluative criteria. These criteria and the indices used for measuring them are described in § 2.3.

Table 5.5 presents the performances and rankings of the quantitative taxonomic methods according to external criterion validity on the iris data base. The external criterion here was the established botanical classification of the iris specimens into three species as specified in Fisher's (1936) paper. Values on three indices of external criterion validity are presented: percentage of concordance with the criterion, Cramér's statistic (whose values take into consideration the size of the cross-classification contingency table and are monotonic with those of other χ^2-related statistics), and a correlation coefficient computed between similarity matrices derived from the grouping produced by a given taxonomic method and the established botanical classification. For each index, magnitudes obtained on each data set and their average as well as a ranking of the taxonomic methods based on their average absolute values, are presented. Average ranks computed by averaging the ranks obtained for each index, and final overall ranks, are also shown. It can be seen that the rankings obtained according to the three indices are considerably similar to each other. The best ranked method was the Rubin-Friedman reallocation technique (9.X). Also well ranked were the complete linkage methods (5.A, 5.B and 5.C), ISODATA (8.X), and NORMIX (10.X). At the other end of the external criterion validity spectrum, the poorest performing methods were *k*-means using correlation coefficients (7.A) and the single linkage methods (4.A, 4.B and 4.C).

Table 5.6 shows the performance and ranking of the quantitative taxonomic methods according to internal criterion validity on the iris data base. Values for the cophenetic correlation coefficient on data sets A and B and their averages, as well as overall ranks, are presented. The best ranked method was single linkage using correlation coefficients (4.A), closely followed by centroid linkage (6.A). Also well ranked were ordinal multidimensional scaling using correlation coefficients (2.A), single linkage using Euclidean and city-block distances (4.B and 4.C), and *k*-means using correlation coefficients (7.A). The poorest ranked method was Chernoff's faces (3.X) followed by ISODATA (8.X), the Rubin-Friedman techique (9.X), NORMIX (10.X) and complete linkage using correlation coefficients (5.A).

Table 5.5

Performance and rankings of quantitative taxonomic methods according to external criterion validity on the iris data base

	Quantitative taxonomic method	% Concordance				Cramér's statistic				Correlation coefficient (between deriv. simil. matrices)				Avg. rank	Overall rank
		Data set A	Data set B	Avg.	Rank	Data set A	Data set B	Avg.	Rank	Data set A	Data set B	Avg.	Rank		
1.A	Q-factor analysis, corr. coef.	73·33	69·33	71·33	12	0·736	0·715	0·7255	13	0·566	0·578	0·572	17	14	14
2.A	Multidimen. scal., corr. coef.	80·00	65·33	72·67	10	0·803	0·707	0·755	12	0·680	0·601	0·641	10	10·67	10
2.B	Multidimen. scal., Euclid. dist.	79·34	81·34	80·34	8	0·791	0·814	0·802	8	0·635	0·690	0·663	8	8	8
2.C	Multidimen. scal., city-bl. dist.	79·34	80·67	80·01	9	0·791	0·807	0·799	9	0·635	0·675	0·655	9	9	9
3.X	Chernoff's faces	86·00	90·00	88·00	6	0·810	0·856	0·833	7	0·648	0·728	0·688	7	6·67	7
4.A	Single linkage, corr. coef.	68·00	65·33	66·67	16·5	0·714	0·707	0·711	16·5	0·613	0·601	0·607	14·5	15·83	16·5
4.B	Single linkage, Euclid. dist.	65·33	68·00	66·67	16·5	0·707	0·714	0·711	16·5	0·601	0·613	0·607	14·5	15·83	16·5
4.C	Single linkage, city-bl. dist.	65·33	69·33	67·33	15	0·707	0·722	0·715	14	0·601	0·601	0·601	16	15	15
5.A	Complete linkage, corr. coef.	93·33	94·67	94·00	2	0·913	0·925	0·919	2	0·816	0·849	0·833	2	2	2
5.B	Complete linkage, Euclid. dist.	90·67	89·33	90·00	3	0·884	0·870	0·877	3	0·757	0·730	0·744	3	3	3
5.C	Complete linkage, city-bl. dist.	88·00	89·33	88·67	5·6	0·857	0·870	0·864	4	0·705	0·730	0·718	4	5·17	4
6.A	Centroid linkage, corr. coef.	68·00	68·00	68·00	14	0·714	0·714	0·714	15	0·613	0·613	0·613	11	13·33	13
7.A	*k*-Means, corr. coef.	53·33	53·33	53·33	18	0·707	0·707	0·707	18	0·468	0·468	0·468	18	18	18
7.B	*k*-Means, Euclid. dist.	90·67	53·33	72·00	11	0·884	0·707	0·976	10·5	0·757	0·468	0·612	12	11·17	11
7.C	*k*-Means, city-bl. dist.	90·67	50·67	70·67	13	0·884	0·707	0·796	10·5	0·757	0·462	0·610	13	12·17	12
8.X	ISODATA	86·67	90·67	88·67	4·5	0·827	0·884	0·856	5	0·673	0·757	0·715	5	4·83	5
9.X	Rubin-Friedman	100·00	100·00	100·00	1	1·000	1·000	1·000	1	1·000	1·000	1·000	1	1	1
10.X	NORMIX	97·33	74·67	86·00	7	0·962	0·717	0·840	6	0·921	0·507	0·714	6	6·33	6

Table 5.6
Performance and rankings of quantitative taxonomic methods according to internal criterion validity on the iris data base

Quantitative taxonomic method		Cophenetic coefficient			Rank
		Data set A	Data set B	Average	
1.A	Q-factor analysis, corr. coef.	0·742	0·808	0·775	7
2.A	Multidimen. scal., corr. coef.	0·750	0·903	0·827	3
2.B	Multidimen. scal., Euclid. dist.	0·741	0·791	0·766	9
2.C	Multidimen. scal., city-bl. dist.	0·749	0·792	0·771	8
3.X	Chernoff's faces	0·502	0·531	0·516	18
4.A	Single linkage, corr. coef.	0·886	0·903	0·895	1
4.B	Single linkage, Euclid. dist.	0·774	0·877	0·826	4
4.C	Single linkage, city-bl. dist.	0·756	0·874	0·815	6
5.A	Complete linkage, corr. coef.	0·625	0·612	0·619	14
5.B	Complete linkage, Euclid. dist.	0·741	0·723	0·732	12
5.C	Complete linkage, city-bl. dist.	0·727	0·731	0·729	13
6.A	Centroid linkage, corr. coef.	0·888	0·891	0·890	2
7.A	*k*-Means, corr. coef.	0·811	0·820	0·816	5
7.B	*k*-Means, Euclid. dist.	0·712	0·772	0·742	10
7.C	*k*-Means, city-bl. dist.	0·722	0·750	0·736	11
8.X	ISODATA	0·602	0·617	0·610	17
9.X	Rubin-Friedman	0·616	0·606	0·611	16
10.X	NORMIX	0·616	0·609	0·613	15

Table 5.7
Performance and ranking of quantitative taxonomic methods according to replicability on the iris data base

Quantitative taxonomic method		Correlation coefficient (between expert-method cross-classification tables)	Rank
1.A	Q-factor analysis, corr. coef.	0·992	5
2.A	Multidimen. scal., corr. coef.	0·852	8
2.B	Multidimen. scal., Euclid. dist.	0·511	10
2.C	Multidimen. scal., city-bl. dist.	0·503	11
3.X	Chernoff's faces	0·973	7
4.A	Single linkage, corr. coef.	0·000	16·5
4.B	Single linkage, Euclid. dist.	0·000	16·5
4.C	Single linkage, city-bl. dist.	0·021	15
5.A	Complete linkage, corr. coef.	0·976	6
5.B	Complete linkage, Euclid. dist.	0·999	2·5
5.C	Complete linkage, city-bl. dist.	0·999	2·5
6.A	Centroid linkage, corr. coef.	0·042	14
7.A	*k*-Means, corr. coef.	—0·421	18
7.B	*k*-Means, Euclid. dist.	0·205	12
7.C	*k*-Means, city-bl. dist.	0·155	13
8.X	ISODATA	0·993	4
9.X	Rubin-Friedman	1·000	1
10.X	NORMIX	0·831	9

Table 5.7 exhibits the performance and ranking of the quantitative taxonomic methods according to replicability on the iris data base. For each taxonomic method, a replicability correlation coefficient computed between corresponding entries of the two expert-method cross-classification tables from data sets A and B, and an overall rank, are presented. The top methods were the Rubin-Friedman reallocation technique (9.X), the complete linkage methods (5.B, 5.C and 5.A), ISODATA (8.X) and Q-factor analysis (1.A). The poorest ranked methods were *k*-means using correlation coefficients (7.A), and the single linkage methods (4.A, 4.B and 4.C).

Table 5.8 presents the rankings of the quantitative taxonomic methods on the iris data base according to each of the three major evaluative criteria (external criterion validity, internal criterion validity and replicability) and across them. The Spearman rank correlation coefficient computed between ranks obtained according to external criterion validity and replicability (our cleanest evaluative criteria), was 0·872. In contrast, the rank correlation coefficients between ranks corresponding to internal criterion validity and

Table 5.8

Ranking of the quantitative taxonomic methods across all three major evaluative criteria (external criterion validity (ECV), internal criterion validity (ICV) and replicability) on the iris data base

Quantitative taxonomic method		ECV rank	ICV rank	Replic. rank	Average rank	Overall rank
1.A	Q-factor analysis, corr. coef.	14	7	5	8·67	6·5
2.A	Multidimen. scal., corr. coef.	10	3	8	7	4
2.B	Multidimen. scal., Euclid. dist.	8	9	10	9	8
2.C	Multidimen. scal., city-bl. dist.	9	8	11	9·33	9
3.X	Chernoff's faces	7	18	7	10·67	13
4.A	Single linkage, corr. coef.	16·5	1	16·5	11·33	14
4.B	Single linkage, Euclid. dist.	16·5	4	16·5	12·33	17
4.C	Single linkage, city-bl. dist.	15	6	15	12	15·5
5.A	Complete linkage, corr. coef.	2	14	6	7·33	5
5.B	Complete linkage, Euclid. dist.	3	12	2·5	5·83	1
5.C	Complete linkage, city-bl. dist.	4	13	2·5	6·5	3
6.A	Centroid linkage, corr. coef.	13	2	14	9·67	10
7.A	*k*-Means, corr. coef.	18	5	18	13·67	18
7.B	*k*-Means, Euclid. dist.	11	10	12	11	12
7.C	*k*-Means, city-bl. dist.	12	11	13	12	15·5
8.X	ISODATA	5	17	4	8·67	6·5
9.X	Rubin-Friedman	1	16	1	6	2
10.X	NORMIX	6	15	9	10	11

Table 5.9
Performance and ranking of the ordinal multidimensional scaling and Chernoff's faces methods according to inter-rater agreement on the iris data base

	Quantitative taxonomic method	% Concordance				Cramér's statistic				Correlation coefficient				Avg. rank	Overall rank
		Data set A	Data set B	Avg.	Rank	Data set A	Data set B	Avg.	Rank	Data set A	Data set B	Avg.	Rank		
2.A	Mutlidimen. scal., corr. coef.	78·67	100·00	89·34	2	0·733	1·000	0·867	1	0·651	1·000	0·826	1	1·33	1
2.B	Multidimen. scal., Euclid. dist.	85·33	78·67	82·00	4	0·776	0·743	0·760	4	0·710	0·642	0·676	4	4·00	4
2.C	Multidimen. scal., city-bl. dist.	85·33	80·00	82·67	3	0·776	0·746	0·761	3	0·710	0·654	0·682	3	3·00	3
3.X	Chernoff's faces	85·33	96·00	90·67	1	0·775	0·942	0·859	2	0·697	0·892	0·795	2	1·67	2

the other two evaluative criteria were quite negative (—0·802 with external criterion validity and —0·706 with replicability). All these three rank correlation coefficients were significantly different from zero at $p < 0{\cdot}01$. Across evaluative criteria, the best ranked taxonomic approaches were the complete linkage methods (5.B, 5.C and 5.A), the Rubin-Friedman technique (9.X) and ordinal multidimensional scaling using correlation coefficients. The poorest ranked methods were k-means using correlation coefficients and city-block distances (7.A and 7.C) and the single linkage methods (4.B, 4.C and 4.A).

Table 5.9 exhibits the performance and ranking of the ordinal multidimensional scaling and Chernoff's faces methods according to inter-rater agreement on the iris data base. These taxonomic methods require the participation of people to complete the clustering process. The evaluation indices are formally similar to those used for assessing external criterion validity. The best performing methods turned out to be ordinal multidimensional scaling using correlation coefficients (2.A) and the Chernoff's faces method (3.X), followed by ordinal multidimensional scaling using city-block and Euclidean distances (2.C and 2.B). However, the ranking does not seem to be conclusive as the differences in absolute performance values among these taxonomic methods were small, especially between the first and second and between the third and fourth ranked methods.

5.5 PERFORMANCE OF RELATIONSHIP MEASURES

The performance of the three relationship measures used in the present study, namely correlation coefficient (A), Euclidean distance (B), and city-block distance (C), was assessed by averaging the performance values of the forms of the ordinal multidimensional scaling (2), single linkage (4), complete linkage (5) and k-means (7) methods which used the corresponding relationship measures. These four taxonomic approaches used all three relationship measures. For example, the performance of the Euclidean distance (B) was assessed by averaging the performance values of 2.B, 4.B, 5.B and 7.B.

Table 5.10 presents the performance and ranking of the three relationship measures on the iris data base, according to external criterion validity, internal criterion validity, replicability and across all three evaluative criteria. The Euclidean distance ranked better than either the correlation coefficient or the city-block distance, but this ordering appears to be tentative and fragile as the differences in absolute performance values among the relationship measures were quite small.

Table 5.10

Performance and ranking of relationship measures according to external criterion validity (ECV), internal criterion validity (ICV) and replicability, as well as across them, on the iris data base

Relationship measure	ECV				ICV		Replicability		Average rank	Overall rank
	% Concordance	Cramér	Corr. coef.	Overall average rank	Cophenetic corr.	Rank	Corr. coef.	Rank		
A Correlation coefficient	71·67	0·773	0·637	3	0·789	1	0·350	3	2·33	2·5
B Euclidean distance	77·25	0·796	0·656	1	0·767	2	0·429	1	1·33	1·0
C City-block distance	76·67	0·793	0·646	2	0·762	3	0·420	2	2·33	2·5

5.6 CLUSTER-BY-CLUSTER AGREEMENT BETWEEN ESTABLISHED BOTANICAL CLASSIFICATION AND QUANTITATIVE TAXONOMIC RESULTS

The grouping produced by each quantitative taxonomic method on each iris data set was cross-tabulated with the established botanical classification of the iris specimens. All the resulting cross-classification tables were summed cell-by-cell in order to assess the rate of recovery of each of the three iris groups through the various quantitative taxonomic methods as well as the major forms of species overlap and misclassification. There were 44 cross-classification tables: 22 for each data set, including two tables for each of the three ordinal multidimensional scaling methods and the Chernoff's faces method as two judges produced separate clusterings for each of these methods. Table 5.11 presents this summed information.

Table 5.11

Cell-by-cell sum of cross-classification tables between established botanical classification and quantitative taxonomic methods on the iris data base

Established botanical groups	Clusters produced by quantitative taxonomic methods			
	Iris setosa	*Iris versicolor*	*Iris virginica*	
Iris setosa	1053	15	32	1100
Iris versicolor	19	764	317	1100
Iris virginica	0	331	769	1100
	1072	1110	1118	3300

Table 5.12 was developed from Table 5.11 to present the percentages of each one of the established iris species groups, as specified by Fisher (1936), classified into the various clusters produced by the quantitative taxonomic methods. An average 96% of the *I. setosa* specimens were clustered as such by the various taxonomic methods. On the other hand, approximately 70% of both the *I. versicolor* and of the *I. virginica* were clustered as such. Most of the remainder of the last two species were cross-clustered between them. The amount of misclassification between *I. setosa* and the other two species groups was very small.

These findings suggest that *I. setosa* is a much more cohesive and distinct species than *I. versicolor* and *I. virginica*, both of which significantly overlap with each other. These results are quite consistent with those of

Table 5.12

Percentage of each one of the established iris species groups classified into the various clusters produced by the quantitative taxonomic methods on the iris data base

Established botanical groups	Clusters produced by quantitative taxonomic methods			
	Iris setosa	*Iris versicolor*	*Iris virginica*	
Iris setosa	95·73	1·36	2·91	100·00
Iris versicolor	1·73	69·45	28·82	100·00
Iris virginica	0·00	30·09	69·91	100·00

Fisher (1936), Kendall (1966), Friedman and Rubin (1967) and Solomon (1971) who, using different classificatory and clustering procedures, were able to recover *I. setosa* specimens very well and had varying degrees of difficulty in recovering *I. versicolor* and *I. virginica* specimens as such, noting a certain amount of overlap between them. In fact, Fisher highlights the genetic reasons for this.

6

Cluster Analysis of Ethnic Populations

6.1 INTRODUCTION

We are living in an increasingly interdependent world, in which there seems to be a greater awareness of the variety of cultures and peoples, and a growing appreciation of their particular contributions to a pluralistic world. With regard to human behavior, there is wide consensus on the powerful role of culture and ethnic identification on the way people live, what they experience and how they relate to each other.

Ethnic populations are defined by their cultural heritage, predominant geographical distribution and some physical characterstics. Migratory patterns, telluric and climatic conditions, social learning and genetic endowment are thought to be the factors underlying the formation of ethnic populations.

The present study examined the cluster organization of ten ethnic populations measured on a set of biochemical genetic variables. Groupings of the ethnic populations were developed through various major quantitative taxonomic methods. Some of the clustering processes and the outcomes obtained on this data base are described in the next section. Then, the systematic performance of the taxonomic methods and the various relationship measures on the ethnic populations are presented; the evaluative criteria used are described in § 2.3. Finally, the levels and areas of agreement and disagreement between the cluster configuration provided by an expert in the field and those generated by the various quantitative taxonomic methods are presented.

6.2 DATA BASE

The ethnic populations data base was developed by Luigi L. Cavalli-Sforza, Professor of Genetics at Stanford University and a widely known expert on the genetics of human populations. The ten ethnic populations studied were the following:

(1) *Eskimo.* From North America, physically characterized by short stature, stocky build, light brown complexion, and broad and flat facial structure.
(2) *Maya.* Amerindian people from the Peninsula of Yucatan.
(3) *English.* People of England, as distinguished from the Scots, Welsh and Irish.
(4) *Italian.* People of Italy.
(5) *Pygmy.* Negroid people of small stature from equatorial Africa.
(6) *Bushman.* Nomadic, ethnically distinct, short-statured people of southern Africa.
(7) *Chinese.* People of China.
(8) *Japanese.* People of Japan.
(9) *Australian aborigine.*
(10) *New Guinean.* People from the large island north of Australia.

Professor Cavalli-Sforza grouped the ten ethnic populations into five pairs: American natives (Eskimos and Mayas), Europeans (English and Italians), Africans (Pygmies and Busmen), Orientals (Chinese and Japanese) and Oceanics (Australian aborigines and New Guineans). These five pairs in fact correspond to the five major continents of the world. However, their geographical centers are not distributed in a uniform or equidistant way. A look at the map shows that the predominant locations of the European and African groups are much closer to each other than to those of the other groups, and the same is true for the Oriental and Oceanic groups.

The ethnic populations were measured on the same set of 58 genetic variables. These variables corresponded to population frequencies of certain alleles including red blood-cell antigens, white blood-cell antigens, plasma proteins and enzymes. Gene frequencies of the ten populations were ranked, and the rank number replaced with the average value of the 1st, 2nd, . . ., 10th individual in a sample of ten from a normal distribution with mean 0 and variance 1.

In order to use these data for studying the stability of cluster configurations generated across sets of variables, the 58 variables were randomly divided into two groups of 29 variables each. In this way, data sets A and B were obtained, each composed of the same ten ethnic populations but measured on different groups of variables. Any expectation of stability of cluster configurations across data sets is based on the fact that the two groups of variables, on the one hand, were relatively large and, on the other hand, were obtained by halving the original group of 58 variables in a

Table 6.1
Data matrix of ten ethnic populations measured on 29 biochemical genetic variables: data set A

Ethnic population	Biochemical genetic variables									
	2	5	6	8	9	10	15	16	17	19
	20	22	28	29	30	31	34	35	36	40
	42	43	46	48	49	52	55	56	57	
American										
Eskimo	—0·372	0·686	—0·239	0·079	—0·093	—0·339	0·070	0·360	—0·890	1·270
	0·070	0·589	1·038	—0·807	0·164	—0·536	—1·433	—0·754	2·685	0·838
	0·401	0·395	1·953	0·620	0·263	0·534	—0·742	—0·337	—0·367	
Maya	0·994	—1·537	0·071	—0·428	0·876	—1·460	1·690	1·270	0·790	—1·000
	0·600	0·693	—1·371	—0·402	—0·180	—0·674	0·316	—1·804	—0·216	0·271
	—0·035	0·475	—3·269	2·557	—0·633	—0·413	0·803	—0·212	0·419	
European										
English	1·327	—1·370	—0·030	—0·222	0·174	0·461	0·220	—0·588	—0·360	0·180
	—0·520	0·969	0·002	—1·418	—0·180	—0·674	0·126	1·395	0·283	—0·644
	0·258	0·026	0·344	0·070	0·670	—0·321	0·715	—0·465	—0·443	
Italian	0·594	—1·077	0·626	—0·145	—0·480	—0·704	—0·070	—0·588	—1·690	0·290
	—1·000	1·842	0·714	—1·590	—0·550	—0·866	—0·753	—0·456	—1·351	—0·090
	—0·121	0·119	0·191	0·546	—1·639	—0·018	0·010	0·026	—0·637	

African										
Pigmy	0·154	2·575	—1·342	—1·176	0·193	0·626	—0·400	—0·588	2·110	1·690
	1·000	—1·847	—2.569	0·842	—1·431	—1·349	—1·324	—2·648	—0·284	0·160
	—0·426	0·185	—1·957	0·278	1·382	—0·793	0·043	0·070	—0·668	
Bushman	—0·103	0·351	—0·965	0·935	—1·519	—0·283	0·140	—0·588	0·890	2·110
	2·110	—2·462	—2·220	—0·140	0·016	—0·597	—1·126	—3·249	0·318	—1·167
	—0·104	—0·617	—1·970	—0·377	1·430	—0·653	—0·378	—0·076	0·222	
Oriental										
Chinese	—1·019	0·540	0·267	0·051	—0·288	0·823	—0·600	1·000	—0·520	—1·450
	0·790	—1·519	—1·236	0·846	1·017	—0·166	0·882	1·051	—0·865	—0·155
	0·930	0·137	0·170	—1·133	0·321	—0·955	—0·207	0·168	—0·043	
Japanese	—1·814	—2·049	0·858	1·055	—0·953	1·153	—0·790	0·790	—0·140	—0·690
	0·690	—1·073	—0·509	—0·129	1·139	—0·118	—0·341	—0·099	—0·496	—0·539
	0·026	0·131	—0·058	—1·362	0·900	—0·422	0·366	0·205	0·722	
Oceanics										
Australian	—1·000	—1·172	—0·352	—0·272	—0·076	2·692	—0·890	—0·588	1·000	—0·327
aborigine	0·220	0·104	—1·608	0·847	0·648	—0·328	—1·842	0·221	—1·521	—0·150
	—0·434	0·566	—1·760	—0·210	—0·334	—0·353	—0·621	—0·205	—0·323	
New	—0·598	1·296	1·940	—0·247	0·068	3·162	1·120	—0·588	1·120	0·600
Guinean	—1·270	—1·430	1·546	0·814	1·410	—0·011	—2·356	1·225	—0·452	—0·361
	0·103	—0·163	—0·063	1·084	—1·155	—0·352	—0·325	0·114	0·514	

Table 6.2
Data matrix of ten ethnic populations measured on 29 biochemical genetic variables: data set B

Ethnic population	Biochemical genetic variables									
	2	5	6	8	9	10	15	16	17	19
	20	22	28	29	30	31	34	35	36	40
	42	43	46	48	49	52	55	56	57	
American										
Eskimo	—0·318	—0·719	—1·285	0·156	2·541	—0·425	—0·849	0·360	0·0	0·898
	—0·291	—1·755	—1·270	—2·274	—2·220	—1·685	—0·452	—0·351	—0·452	0·388
	0·606	1·539	—1·052	1·123	—0·164	—0·181	—0·326	—1·367	—0·358	
Maya	2·058	0·173	—2·195	0·674	1·183	0·517	—0·070	0·220	—0·690	1·448
	0·101	—0·798	0·600	0·478	—0·807	—1·864	—2·865	0·054	—1·266	0·584
	—0·289	0·018	—2·166	1·488	—1·496	—0·015	—0·435	0·330	—0·258	
European										
English	—0·385	—0·288	1·288	—0·941	—1·124	—0·343	0·890	—0·600	—0·890	0·283
	0·026	0·206	—1·000	—1·591	—0·109	2·540	—0·793	—0·525	0·649	—0·092
	0·561	—1·875	1·523	0·679	0·068	—0·587	0·103	0·377	0·266	
Italian	—1·336	—0·502	0·755	—0·165	—0·884	—0·203	0·440	—0·790	—0·600	—1·542
	—0·544	0·662	—0·363	—1·161	0·974	2·079	0·156	1·268	0·910	0·178
	0·623	—2·254	0·967	—1·648	—0·282	—0·937	0·105	0·466	—0·141	

African										
Pigmy	1·118	—0·702	1·481	—1·104	0·672	0·139	1·690	—1·450	—0·255	—1·419
	0·148	—0·790	1·450	—1·650	1·905	1·919	—0·433	0·907	0·216	—0·473
	—0·688	—1·540	0·803	0·985	0·114	0·328	—0·039	—0·388	—0·200	
Bushman	0·078	0·692	3·609	—1·126	0·796	1·139	—0·849	—2·110	1·103	—2·769
	0·354	0·399	0·890	1·869	—0·080	1·875	—0·866	—0·243	—1·067	0·277
	—0·054	—1·416	0·358	0·470	—0·378	0·142	—0·399	—0·798	—0·354	
Oriental										
Chinese	—0·693	0·524	—1·403	0·129	1·994	0·287	—0·849	1·750	0·290	—0·260
	0·640	0·212	—0·690	0·894	—0·584	—0·482	1·208	0·884	—0·624	0·196
	—0·973	1·490	1·169	—1·241	—0·299	1·712	—0·691	—0·082	0·299	
Japanese	—1·901	—0·289	—0·410	—0·649	0·909	0·282	—0·849	0·790	1·103	1·070
	0·855	0·266	0·220	1·259	—0·395	—2·591	0·926	—0·674	1·418	1·085
	—0·526	1·591	—0·608	—1·168	—0·758	—0·873	1·367	0·206	0·155	
Oceanics										
Australian aborigine	—0·156	—2·956	—2·792	0·245	0·436	—0·019	—0·849	1·270	1·103	2·005
	0·662	0·069	1·690	3·138	0·575	—2·214	2·117	—1·007	—0·954	—0·578
	—0·333	4·153	2·973	—0·334	—8·084	0·069	0·787	0·720	—0·723	
New Guinean	0·276	0·292	—1·968	—1·127	—0·994	—0·167	—0·849	0·890	1·103	0·046
	0·506	—0·405	—2·110	1·099	—0·478	—2·264	2·725	—0·657	—1·730	—0·451
	0·509	3·166	3·139	—1·085	1·440	—0·646	—1·367	0·290	0·123	

Table 6.3

Means and standard deviations of five groups of ethnic populations on 29 biochemical genetic variables of data sets A and B

Data set A

Ethnic groups		Var 2	Var 5	Var 6	Var 8	Var 9	Var 10	Var 15	Var 16	Var 17	Var 19	Var 20	Var 22	Var 28	Var 29	Var 30
Amer.	Mean	0·311	—0·425	—0·084	—0·174	0·391	—0·899	0·880	0·815	—0·050	0·135	0·335	0·641	—0·167	—0·604	—0·008
(*N*=2)	SD	0·966	1·572	0·219	0·359	0·685	0·793	1·146	0·643	1·188	1·605	0·375	0·074	1·703	0·286	0·243
Euro.	Mean	0·960	—1·223	0·298	—0·183	—0·153	—0·122	0·075	—0·588	—1·025	0·235	—0·760	1·405	0·358	—1·504	—0·365
(*N*=2)	SD	0·518	0·207	0·464	0·054	0·462	0·824	0·205	0·001	0·940	0·078	0·339	0·617	0·503	0·122	0·262
Afri.	Mean	0·025	1·463	—1·153	—0·120	—0·663	0·171	—0·130	—0·588	1·500	1·900	1·555	—2·154	—2·394	0·351	—0·707
(*N*=2)	SD	0·182	1·573	0·267	1·493	1·211	0·643	0·382	0·001	0·863	0·297	0·785	0·435	0·247	0·694	1·023
Orie.	Mean	—1·416	—0·754	0·562	0·553	—0·620	0·988	—0·695	0·895	—0·330	—1·070	0·740	—1·296	—0·872	0·359	1·078
(*N*=2)	SD	0·562	1·831	0·418	0·710	0·470	0·233	0·134	0·148	0·269	0·537	0·071	0·315	0·514	0·689	0·086
Ocea.	Mean	—0·799	0·062	0·794	—0·259	—0·004	2·927	0·115	—0·588	1·060	0·137	—0·525	—0·663	—0·031	0·830	1·029
(*N*=2)	SD	0·284	1·745	1·621	0·018	0·102	0·332	1·421	0·001	0·085	0·655	1·054	1·085	2·230	0·023	0·539

		Var 31	Var 34	Var 35	Var 36	Var 40	Var 42	Var 43	Var 46	Var 48	Var 49	Var 52	Var 55	Var 56	Var 57
Amer.	Mean	—0·605	—0·558	—1·279	1·234	0·554	0·183	0·435	—0·658	1·588	—0·185	0·060	0·030	—0·274	0·026
(*N*=2)	SD	0·098	1·237	0·742	2·051	0·401	0·308	0·057	3·693	1·370	0·634	0·670	1·092	0·088	0·556
Euro.	Mean	—0·770	—0·313	0·469	—0·534	—0·367	0·068	0·072	0·267	0·308	—0·484	—0·169	0·362	—0·219	—0·540
(*N*=2)	SD	0·136	0·622	1·309	1·155	0·392	0·268	0·066	0·108	0·337	1·633	0·214	0·499	0·347	0·137
Afri.	Mean	—0·973	—1·225	—2·948	0·017	—0·503	—0·265	—0·216	—1·963	—0·049	1·406	—0·723	—0·167	—0·003	—0·223
(*N*=2)	SD	0·532	0·140	0·425	0·426	0·938	0·228	0·567	0·009	0·463	0·034	0·099	0·298	0·103	0·629
Orie.	Mean	—0·142	0·271	0·476	—0·680	—0·347	0·478	0·134	0·056	—1·247	0·610	—0·688	0·080	0·186	0·339
(*N*=2)	SD	0·034	0·865	0·813	0·261	0·272	0·639	0·004	0·161	0·162	0·409	0·377	0·405	0·026	0·541
Ocea.	Mean	—0·169	—2·099	0·723	—0·986	—0·255	—0·165	0·201	—0·911	0·437	—0·744	—0·352	—0·473	—0·046	0·095
(*N*=2)	SD	0·224	0·363	0·710	0·756	0·149	0·380	0·515	1·200	0·915	0·581	0·001	0·209	0·226	0·592

Data set B

Ethnic groups		Var 1	Var 3	Var 4	Var 7	Var 11	Var 12	Var 13	Var 14	Var 18	Var 21	Var 23	Var 24	Var 25	Var 26	Var 27
Amer.	Mean	0·870	—0·273	—1·740	0·415	1·862	0·046	—0·459	0·290	—0·345	1·173	—0·095	—1·276	—0·335	—0·898	—1·513
($N=2$)	SD	1·680	0·631	0·643	0·366	0·960	0·666	0·551	0·099	0·488	0·389	0·277	0·677	1·322	1·946	0·999
Euro.	Mean	—0·860	—0·395	1·021	—0·553	—1·004	—0·273	0·665	—0·695	—0·745	—0·629	—0·259	0·434	—0·681	—1·376	0·433
($N=2$)	SD	0·672	0·151	0·377	0·549	0·170	0·099	0·318	0·134	0·205	1·290	0·403	0·322	0·450	0·304	0·766
Afri.	Mean	0·598	—0·005	2·545	—1·115	0·734	0·639	0·420	—1·780	0·424	—2·094	0·251	—0·195	1·170	0·109	0·912
($N=2$)	SD	0·735	0·986	1·505	0·016	0·088	0·707	1·795	0·467	0·960	0·955	0·146	0·841	0·396	2·488	i·404
Orie.	Mean	—1·297	0·118	—0·906	—0·260	1·451	0·284	—0·849	1·270	0·696	0·405	0·747	0·239	—0·235	1·076	—0·489
($N=2$)	SD	0·854	0·575	0·702	0·550	0·767	0·004	0·000	0·679	0·575	0·940	0·152	0·038	0·643	0·258	0·134
Ocea.	Mean	0·060	—1·332	—2·380	—0·441	—0·279	—0·093	—0·849	1·080	1·103	1·025	0·584	—0·168	—0·210	2·118	0·049
($N=2$)	SD	0·305	2·297	0·583	0·970	1·011	0·105	0·000	0·269	0·000	1·385	0·110	0·335	2·687	1·442	0·745

		Var 32	Var 33	Var 37	Var 38	Var 39	Var 41	Var 44	Var 45	Var 47	Var 50	Var 51	Var 53	Var 54	Var 58
Amer.	Mean	—1·774	—1·658	—0·148	—0·859	0·486	0·158	0·778	—1·609	1·305	—0·830	—0·098	—0·380	—0·518	—0·308
($N=2$)	SD	0·127	1·706	0·286	0·576	0·139	0·633	1·076	0·788	0·258	0·942	0·117	0·077	1·200	0·071
Euro.	Mean	2·309	—0·318	0·371	0·779	0·043	0·592	—2·064	1·245	—0·484	—0·107	—0·762	0·104	0·421	0·063
($N=2$)	SD	0·326	0·671	1·268	0·185	0·191	0·044	0·268	0·393	1·645	0·247	0·247	0·001	0·063	0·288
Afri.	Mean	1·897	—0·649	0·332	—0·425	—0·098	—0·371	—1·478	0·580	0·727	—0·132	0·235	—0·219	—0·593	—0·277
($N=2$)	SD	0·031	0·306	0·813	0·907	0·530	0·448	0·088	0·315	0·0364	0·348	0·132	0·255	0·290	0·109
Orie.	Mean	—1·536	1·067	0·105	0·397	0·640	—0·749	1·540	0·280	—1·204	—0·528	0·419	0·338	0·062	0·227
($N=2$)	SD	1·491	0·199	1·102	1·444	0·629	0·316	0·071	1·257	0·052	0·325	1·828	1·455	0·204	0·102
Ocea.	Mean	—2·239	2·421	—0·832	—1·342	—0·514	0·088	3·659	3·056	—0·709	0·678	—0·289	—0·290	0·505	—0·300
($N=2$)	SD	0·036	0·430	0·247	0·549	0·090	0·595	0·698	0·117	0·531	1·078	0·506	1·523	0·304	0·598

random way rather than reflecting conceptual principles or systematic biases.

Tables 6.1 and 6.2 present data matrices constituted by the values of the same ten ethnic populations on each of the 29 genetic variables of data sets A and B, respectively.

Table 6.3 presents the means and standard deviations of the five pairs of ethnic populations on the 29 variables of data sets A and B. Figures 6.1 and 6.2 show, for data sets A and B respectively, the mean graphical profiles of the five pairs of ethnic populations. The relatively large number of variables in each data set make it difficult to appreciate interprofile differences.

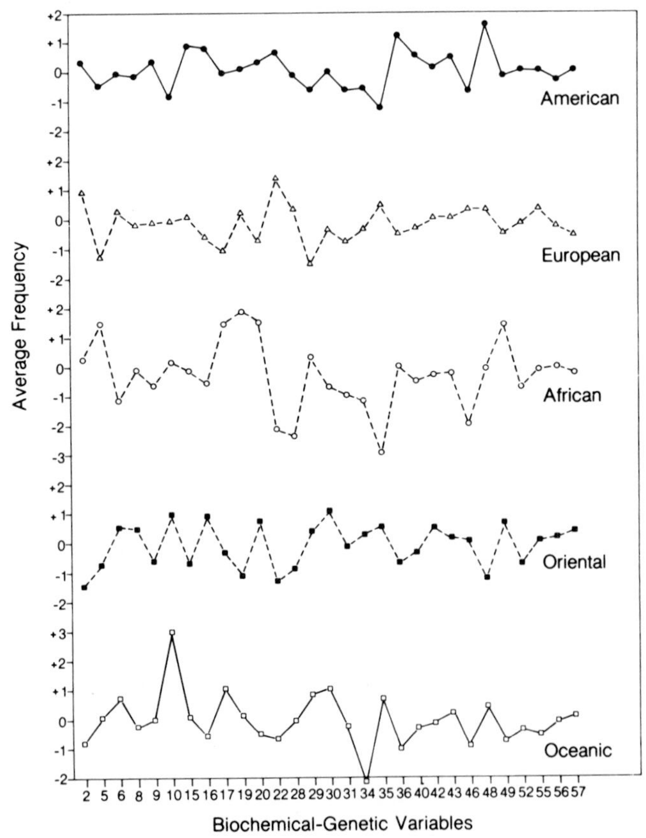

Fig. 6.1. Profiles of five population groups on data set A.

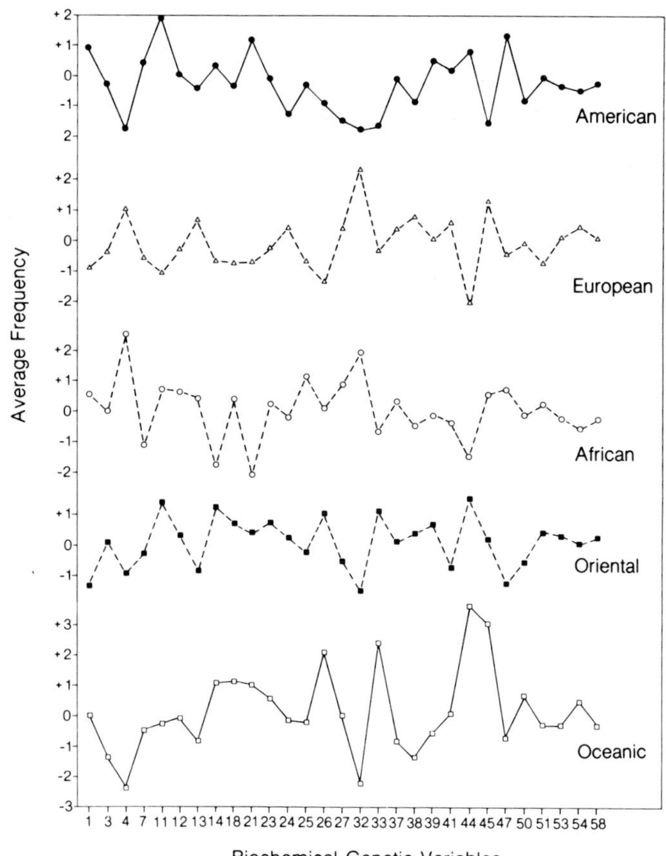

Fig. 6.2 Profiles of five population groups on data set B.

6.3 ILLUSTRATIVE CLUSTER ANALYTIC RESULTS

The various quantitative taxonomic methods listed in Table 2.1, except the Rubin-Friedman technique and NORMAP/NORMIX, were applied to the ethnic population data sets. The two methods mentioned above were not applied on this data base because the expert-based clusters included only two entities each. The clustering outcome obtained by applying ordinal multidimensional scaling and the complete linkage method on both data sets A and B will be presented first, as was done in the case of the data bases studied in Chapters 3, 4 and 5. This will allow a comparative appreciation of the application of two important taxonomic methods across fields.

Additionally, the results of the cluster analysis of data set A through the single linkage method and of data set B through the centroid linkage method, will be exhibited. In this way, with the six other taxonomic methods illustrated on the previous three data bases, all our ten basic clustering approaches will be covered.

Figures 6.3 and 6.4 show two-dimensional ordinal multidimensional scaling representations, using correlation coefficients, of the ten ethnic populations for data sets A and B (which have different sets of variables), respectively. For data set A, ethnic populations 3 and 4 (Europeans), 5 and 6 (Africans), and 7 and 8 (Orientals) appear respectively clustered with or adjacent to each other. The two judges who visually examined this configuration put together each one of these pairs, but sometimes with the addition of some nearby point. On the other hand, populations 1 and 2 (native Americans), and 9 and 10 (Oceanic) are relatively distant from each other, not forming visualizable clustered pairs. The data set B configuration turned out to have quite a low stress value (0·100) (especially low in comparison with the stress of 0·367 obtained in data set A), indicating that the rank of interpoint distances on this two-dimensional configuration does not differ much from the rank of interpoint distances on the input data. Here, as in the case of data set A, ethnic populations 3 (English) and 4 (Italian) appear quite close to each other, forming a European pair. The other continental pairs defined by the expert do not form distinct clusters, and are not even closer to each other than to any other population.

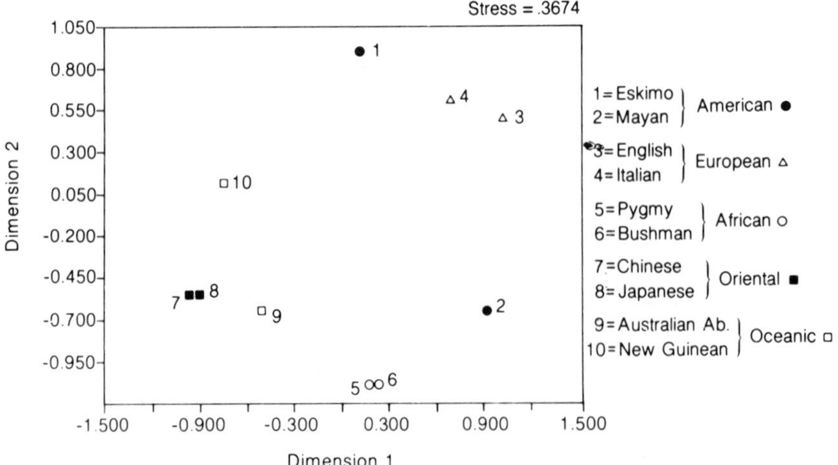

Fig. 6.3 Ordinal multidimensional scaling representation, using correlation coefficients, of the ten ethnic populations of data set A.

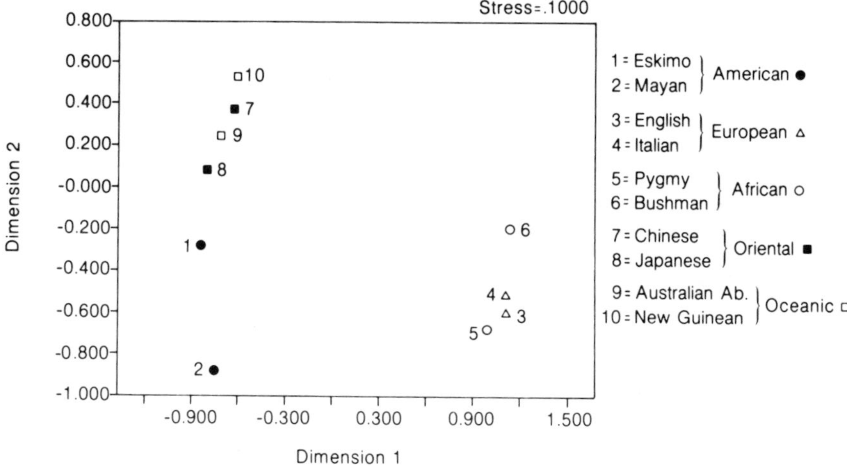

Fig. 6.4 Ordinal multidimensional scaling representation, using correlation coefficients, of the ten ethnic populations of data set B.

However, it is possible to detect a cluster on the lower right corner of the field, formed by the two European populations (3 and 4) and the two African populations (5 and 6), as well as, somewhat less clearly, a cluster on the upper left corner constituted by the Oriental (7 and 8) and Oceanic (9 and 10) populations. The two judges who visually examined this plot consistently put the Oriental and Oceanic populations together in one cluster. The two continents represented on each corner are, in fact, geographically closer to each other than to any of the other three continents.

Figures 6.5 and 6.6 present dendrograms depicting the complete linkage agglomerative cluster anlysis, using correlation coefficients, of the ten ethnic populations in data sets A and B, respectively, In data set A, ethnic populations 5 (Pygmy) and 6 (Bushman) clustered first, forming an African pair. Slightly later in the process, populations 7 and 8 (Orientals), 9 and 10 (Oceanics), and 3 and 4 (Europeans), clustered, forming the indicated continental clusters in line with the expert's classification. Much later, population 1 (Eskimo) joined the already established European cluster, and population 2 (Mayan) joined the African cluster. In data set B, almost at the same similarity level, ethnic populations 9 (Australian aborigine) and 10 (New Guinean) formed an Oceanic cluster, and 3 (English) and 4 (Italian) formed a European pair. Later in the agglomerative process, and in a sequential way, populations 7 (Chinese) and 8 (Japanese) joined the previously established Oceanic cluster, and populations 5 (Pygmy) and 6

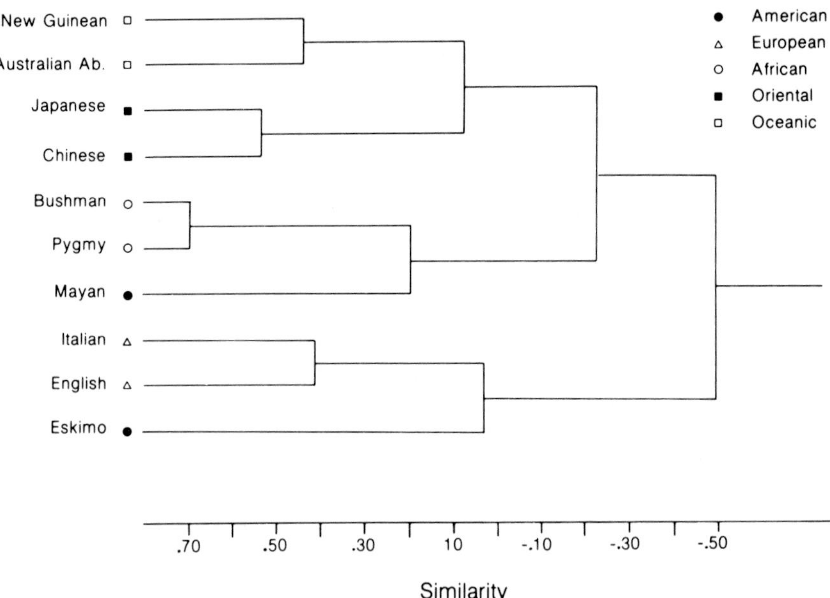

Fig. 6.5 Dendrogram representing a complete linkage cluster analysis, using correlation coefficients, of the ten ethnic populations of data set A.

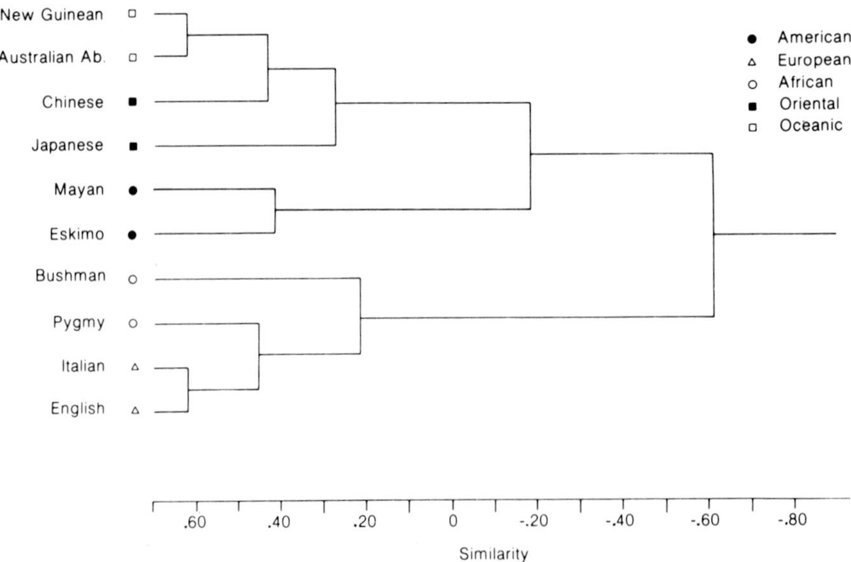

Fig. 6.6 Dendrogram representing a complete linkage cluster analysis, using correlation coefficients, of the ten ethnic populations of data set B.

(Bushman) joined the already constituted European cluster. As indicated for the ordinal multidimensional scaling representation of data set B, the European-African cluster and the Oriental-Oceanic cluster, which appeared here at about similarity level 0·200, include continents which are geographically closer to each other than to any of the other three major continents. At an intermediate point in the agglomerative process, populations 1 (Eskimo) and 2 (Mayan) clustered with each other and, very late in the process, joined the Oriental-Oceanic cluster.

Figure 6.7 shows a dendrogram displaying the single linkage cluster analysis, using correlation coefficients, of the ten ethnic populations in data set A. The general agglomerative process obtained through this method is similar to that for complete linkage shown in Figs 6.5 and 6.6. The main methodological difference refers to the definition of distance between clusters which, as discussed in § 2.1, is the distance between the furthest apart members of the clusters in the complete linkage method, while in the single linkage method it is the distance between the closest two members, one from each cluster. As can be seen, the clustering process and outcome

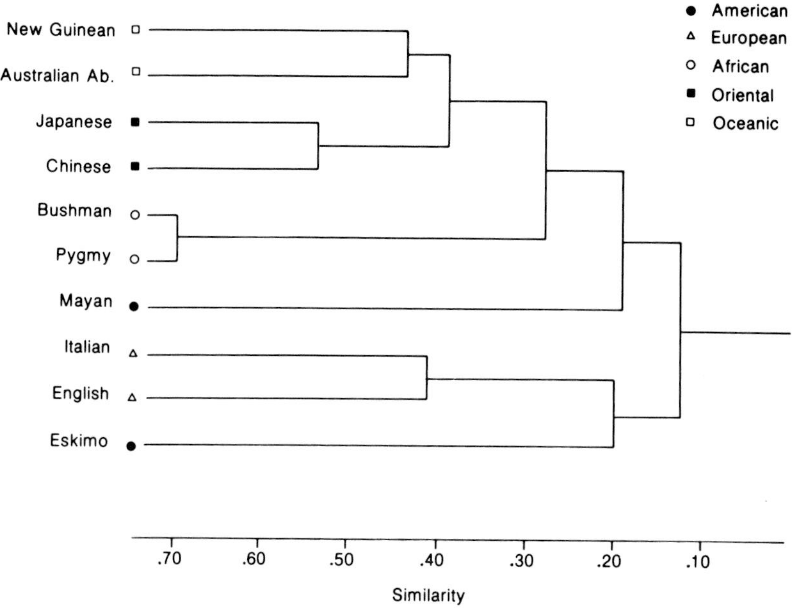

Fig. 6.7 Dendrogram representing a single linkage cluster analysis, using correlations coefficients, of the ten ethnic populations of data set A.

obtained through the single linkage method here, are quite comparable to those obtained through complete linkage in data set A and shown in Fig. 6.5.

Figure 6.8 exhibits a dendrogram reflecting the centroid linkage cluster analysis, using correlation coefficients, of the ten ethnic populations in data set B. The centroid linkage method is hierarchical and agglomerative, as are the complete linkage and single linkage methods. The main difference among these methods refers again to the definition of distance between clusters, which in the case of the centroid linkage method is the distance between cluster centroids. The clustering outcome obtained by this method was very similar to that obtained by the complete linkage method in data set B.

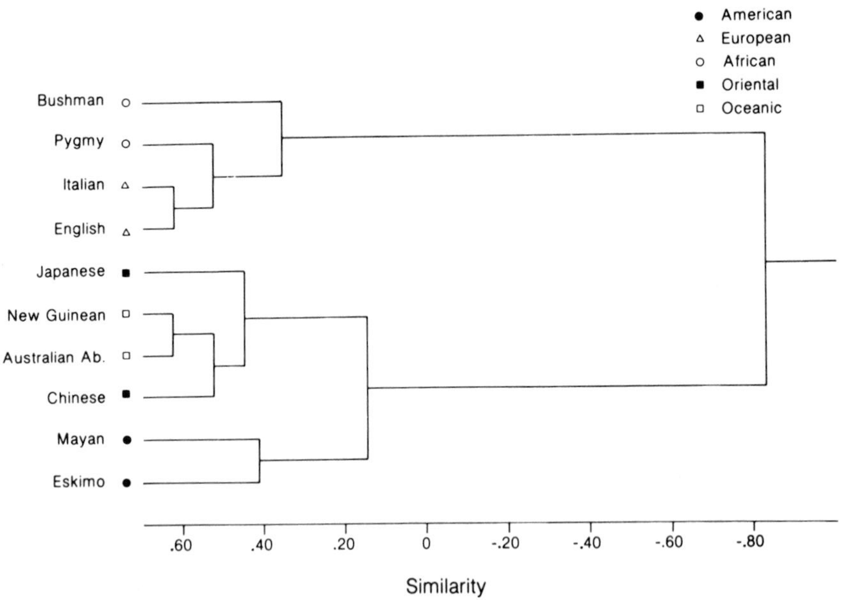

Fig. 6.8 Dendrogram representing a centroid linkage cluster analysis, using correlation coefficients, of the ten ethnic populations of data set B.

One general comment about the cluster configurations of the ten ethnic populations exhibited in this section is that those obtained in data sets A and B have some similarities but also some considerable differences, perhaps more marked differences than in the case of previous data bases. This undoubtedly has to do, at least in part, with the fact that, in contrast to data sets studied in the previous three chapters, here the group of variables used in data set A was different from that used in data set B, although they were relatively large in size and obtained by random (rather than conceptual or systematic bias) division of the original set of 58 variables. In data set A, four ethnic population pairs (European, African, Oriental and Oceanic) of the five representing major continents and classified as such by the expert, were recovered by most methods. In data set B, three of the five pairs were recovered (i.e. native Americans, Europeans and Oceanics), but a meaningful higher-order structure appeared. This was mainly constituted by a European-African cluster and an Oriental-Oceanic cluster, which makes sense from a geographical proximity viewpoint. In data set A, the higher-order Oriental-Oceanic cluster was also detectable.

A word of caution is given here in order to prevent generalizations from the high comparability of the configurations obtained through the complete, single and centroid linkage methods in this data base. The comparability observed here is probably based on the distinctness of the cluster structure of this data base and the small number of entities involved. Cluster analysis of the other data bases showed that although the complete and centroid linkage methods tended to yield similar configurations, these were frequently different from those produced by the single linkage method. The complete and centroid linkage methods tend to be best for recovering circular or spherical clusters while the single linkage method tends to do best with chain-like, serpentine clusters, as shown by Everitt (1974) using artificial data.

6.4 PERFORMANCE OF QUANTITATIVE TAXONOMIC METHODS

The performance of the quantitative taxonomic methods on the ethnic populations data base was evaluated according to external criterion validity, internal criterion validity, replicability and, for ordinal multidimensional scaling and Chernoff's faces, interjudge agreement, as well as across the first three evaluative criteria. These criteria and the indices used for measuring them are described in § 2.3.

The performance and rankings of the quantitative taxonomic methods according to external criterion validity on the ethnic populations data base are presented in Table 6.4. The external criterion here was the grouping by the expert (Professor L. L. Cavalli-Sforza) of the ten ethnic populations into five pairs, which corresponded to the five major geographical continents. Values on three indices of external criterion validity are presented. These are: first, percentage of concordance with the external criterion; second, Cramér's statistic (whose values take into consideration the size of the cross-classification contingency table and were monotonic with the values of other χ^2-related statistics); and third, a correlation coefficient computed

Table 6.4

Performance and rankings of quantitative taxonomic methods according to external criterion validity on the ethnic populations data base

Quantitative taxonomic method		% Concordance				Cramér's statistic				Correlation coefficient				Avg. rank	Overall rank
		Data set A	Data set B	Avg.	Rank	Data set A	Data set B	Avg.	Rank	Data set A	Data set B	Avg.	Rank		
1.A	Q-factor analysis, corr. coef.	70·00	70·00	70·00	9·5	0·866	0·764	0·815	10	0·575	0·277	0·426	11	10·17	9
2.A	Multidimen. scal., corr. coef.	80·00	60·00	70·00	9·5	0·816	0·764	0·790	12	0·434	0·354	0·394	12	11·17	12
2.B	Multidimen. scal., Euclid. dist.	80·00	80·00	80·00	5·5	0·816	0·890	0·853	5	0·434	0·634	0·534	5	5·17	5
2.C	Multidimen. scal., city-bl. dist.	80·00	90·00	85·00	2·5	0·816	0·913	0·865	2·5	0·434	0·693	0·564	4	3	4
3.X	Chernoff's faces	65·00	70·00	67·50	11	0·750	0·790	0·770	13	0·258	0·356	0·307	14	12·67	13
4.A	Single linkage, corr. coef.	70·00	60·00	65·00	13	0·866	0·764	0·815	10	0·575	0·354	0·465	8·5	10·5	10·5
4.B	Single linkage, Euclid. dist.	80·00	70·00	75·00	7·5	0·816	0·866	0·841	6·5	0·437	0·575	0·506	6·5	6·83	6·5
4.C	Single linkage, city-bl. dist.	80·00	70·00	75·00	7·5	0·816	0·866	0·841	6·5	0·437	0·575	0·506	6·5	6·83	6·5
5.A	Complete linkage, corr. coef.	90·00	80·00	85·00	2·5	0·913	0·816	0·865	2·5	0·693	0·437	0·565	2	2·33	2
5.B	Complete linkage, Euclid. dist.	80·00	90·00	85·00	2·5	0·816	0·913	0·865	2·5	0·437	0·693	0·565	2	2·33	2
5.C	Complete linkage, city-bl. dist.	80·00	90·00	85·00	2·5	0·816	0·913	0·865	2·5	0·437	0·693	0·565	2	2·33	2
6.A	Centroid linkage, corr. coef.	70·00	60·00	65·00	13	0·866	0·764	0·815	10	0·575	0·354	0·465	8·5	10·5	10·5
7.A	*k*-Means, corr. coef.	80·00	80·00	80·00	5·5	0·816	0·816	0·816	8	0·437	0·437	0·437	10	7·83	8
7.B	*k*-Means, Euclid. dist.	70·00	50·00	60·00	15	0·764	0·707	0·736	15	0·277	0·206	0·242	15	15	15
7.C	*k*-Means, city-bl. dist.	80·00	50·00	65·00	13	0·816	0·707	0·762	14	0·437	0·206	0·322	13	13·33	14
8.X	ISODATA	60·00	50·00	55·00	16	0·775	0·677	0·726	16	0·292	0·050	0·171	16	16	16

between similarity matrices derived from the grouping produced by a given taxonomic method and from the classification established by the expert. For each index, magnitudes obtained on each data set and their average values across data sets A and B as well as a ranking of the taxonomic methods based on these average absolute values, are presented. Finally, average ranks computed by averaging those obtained for each index, and then overall ranks, are exhibited. The rankings of the taxonomic methods, obtained according to the three indices, are comparable to each other. Clearly, the best ranked methods were complete linkage using correlation coefficients, Euclidean distances and city-block distances (5.A, 5.B and 5.C). Also well ranked was ordinal multidimensional scaling using city-block and Euclidean distances (2.C and 2.B). The poorest performing method was ISODATA (8.X), followed by *k*-means using Euclidean and city-block distances (7.B and 7.C).

Table 6.5 shows the performance and ranking of the quantitative taxonomic methods according to internal criterion validity on the ethnic populations data base. Values for the cophenetic correlation coefficient on data sets A and B and their average, as well as overall ranks, are presented. The best ranked methods were single linkage (4.B, 4.C and 4.A), complete linkage using city-block distances (5.C), ordinal multidimensional scaling using city-block and Euclidean distances (2.B and 2.C), and centroid linkage (6.A). The poorest ranked methods were Q-factor analysis (1.A), ISODATA (8.X) and Chernoff's faces (3.X).

Table 6.5
Performance and ranking of quantitative taxonomic methods according to internal criterion validity on the ethnic populations data base

Quantitative taxonomic method		Cophenetic correlation			Rank
		Data set	Data	Average	
1.A	Q-factor analysis, corr. coef.	−0·029	−0·217	−0·123	16
2.A	Multidimen. scal., corr. coef.	0·599	0·707	0·653	9
2.B	Multidimen. scal., Euclid. dist.	0·726	0·639	0·682	5
2.C	Multidimen. scal., city-bl. dist.	0·724	0·641	0·683	3·5
3.X	Chernoff's faces	0·282	0·242	0·262	14
4.A	Single linkage corr. coef.	0·654	0·707	0·681	6·5
4.B	Single linkage Euclid. dist.	0·726	0·659	0·693	1
4.C	Single linkage, city-bl. dist.	0·724	0·648	0·686	2
5.A	Complete linkage, corr. coef.	0·624	0·645	0·635	10
5.B	Complete linkage, Euclid. dist.	0·726	0·618	0·672	8
5.C	Complete linkage, city-bl. dist.	0·724	0·641	0·683	3·5
6.A	Centroid linkage, corr. coef.	0·654	0·707	0·681	6·5
7.A	*k*-Means, corr. coef.	0·541	0·333	0·437	13
7.B	*k*-Means, Euclid. dist.	0·535	0·607	0·571	12
7.C	*k*-Means, city-bl. dist.	0·582	0·594	0·588	11
8.X	ISODATA	0·219	−0·044	0·088	15

Table 6.6 shows the performance and ranking of the 16 quantitative taxonomic methods evaluated on the ethnic populations data base according to stability of cluster configurations across two groups of variables. As mentioned before, any expectation of stability here was based on the two groups of variables being relatively large and their having been obtained by random rather than by biased division of the original set of 58 variables. For each taxonomic method, a stability or replicability correlation coefficient computed between corresponding entries of the two expert-method cross-classification tables from data sets A and B, and an overall rank, are presented. The top ranked method was *k*-means using correlation coefficients (7.A), followed by single linkage using correlation coefficients (4.A) and centroid linkage (6.A). The poorest ranked method was Q-factor analysis (1.A), followed by *k*-means using city-block and Euclidean distances (7.C and 7.B). Thus, there was a considerable difference in performance between *k*-means using correlation coefficients and *k*-means using either Euclidean or city-block distances. The same contrast regarding relationship measures used was noted for the single linkage approach.

The rankings of the quantitative taxonomic methods on the ethnic population data base, according to each of the three major evaluative

Table 6.6

Performance and ranking of quantitative taxonomic methods according to stability on the ethnic populations data base

Quantitative taxonomic method		Correlation coefficient (between expert-method cross-classification tables)	Rank
1.A	Q-factor analysis, corr. coef.	0·423	16
2.A	Multidimen. scal., corr. coef.	0·750	8
2.B	Multidimen. scal., Euclid. dist.	0·656	10
2.C	Multidimen. scal., city-bl. dist.	0·772	5·5
3.X	Chernoff's faces	0·528	13
4.A	Single linkage, corr. coef.	0·926	2·5
4.B	Single linkage, Euclid. dist.	0·540	11·5
4.C	Single linkage, city-bl. dist.	0·540	11·5
5.A	Complete linkage, corr. coef.	0·772	5·5
5.B	Complete linkage, Euclid. dist.	0·772	5·5
5.C	Complete linkage, city-bl. dist.	0·772	5·5
6.A	Centroid linkage, corr. coef.	0·926	2·5
7.A	*k*-Means, corr. coef.	1·000	1
7.B	*k*-Means, Euclid. dist.	0·500	14
7.C	*k*-Means, city-bl. dist.	0·456	15
8.X	ISODATA	0·730	9

criteria (external criterion validity, internal criterion validity and stability or replicability) and across them, are presented in Table 6.7. A certain degree of variability was noted among the rankings obtained according to the three evaluative criteria. The Spearman rank correlation coefficients of the ranking for external criteria validity with that for internal criterion validity was 0·576, and with that for stability was 0·401. The rank correlation coefficient between the rankings for internal criterion validity and stability was 0·274. Only the first rank correlation coefficient was statistically significantly different from zero. Across the three major evaluative criteria, the top overall ranks were obtained by the complete linkage methods (5.C, 5.B and 5.A) and ordinal multidimensional scaling using city-block distances (2.C). At the other end of the spectrum, the poorest ranked methods were Q-factor analysis (1.A), *k*-means using Euclidean and city-block distances (7.B and 7.C), ISODATA (8.X) and Chernoff's faces (3.X).

Table 6.8 presents the performance and rankings of the ordinal multidimensional scaling and Chernoff's faces methods according to inter-rater agreement on the ethnic populations data base. In each case, two judges participated to complete the clustering process. The evaluation indices were formally similar to those used for assessing external criterion validity. Ordinal multidimensional scaling using correlation coefficients and city-block and Euclidean distances (2.A, 2.C and 2.B) consistently performed better than the Chernoff's faces method (3.X).

Table 6.7

Ranking of the quantitative taxonomic methods across all three major evaluative criteria (external criterion validity (ECV), internal criterion validity (ICV) and stability) on the ethnic populations data base

Quantitative taxonomic method		ECV rank	ICV rank	Stability rank	Average rank	Overall rank
1.A	Q-factor analysis, corr. coef.	9	16	16	13·67	15·5
2.A	Multidimen. scal., corr. coef.	12	9	8	9·67	11
2.B	Multidimen. scal., Euclid. dist.	5	5	10	6·67	8·5
2.C	Multidimen. scal., city-bl. dist.	4	3·5	5·5	4·33	2
3.X	Chernoff's faces	13	14	13	13·33	13
4.A	Single linkage, corr. coef.	10·5	6·5	2·5	6·5	6·5
4.B	Single linkage, Euclid. dist.	6·5	1	11·5	6·33	5
4.C	Single linkage, city-bl. dist.	6·5	2	11·5	6·67	8·5
5.A	Complete linkage, corr. coef.	2	10	5·5	5·83	4
5.B	Complete linkage, Euclid. dist.	2	8	5·5	5·17	3
5.C	Complete linkage, city-bl. dist.	2	3·5	5·5	3·67	1
6.A	Centroid linkage, corr. coef.	10·5	6·5	2·5	6·5	6·5
7.A	*k*-Means, corr. coef.	8	13	1	7·33	10
7.B	*k*-Means, Euclid. dist.	15	12	14	13·67	15·5
7.C	*k*-Means, city-bl. dist.	14	11	15	13·33	13
8.X	ISODATA	16	15	9	13·33	13

Table 6.8
Performance and ranking of the ordinal multidimensional scaling and Chernoff's faces methods according to inter-rater agreement on the ethnic populations data base

Quantitative taxonomic method	% Concordance				Cramér's statistic				Correlation coefficient				Avg. rank	Overall rank
	Data set A	Data set B	Avg.	Rank	Data set A	Data set B	Avg.	Rank	Data set A	Data set B	Avg.	Rank		
2.A Multidimen. scal., corr. coef.	100·00	100·00	100·00	1·5	1·000	1·000	1·000	1·5	1·000	1·000	1·000	1·5	1·5	1·5
2.B Multidimen. scal., Euclid. dist.	100·00	80·00	90·00	3	1·000	0·866	0·933	3	1·000	0·672	0·836	3	3	3
2.C Multidimen. scal., city-bl. dist.	100·00	100·00	100·00	1·5	1·000	1·000	1·000	1·5	1·000	1·000	1·000	1·5	1·5	1·5
3.X Chernoff's faces	60·00	60·00	60·00	4	0·726	0·692	0·727	4	0·192	0·192	0·192	4	4	4

Table 6.9
Performance and rankings of relationship measures according to external criterion validity (ECV), internal criterion validity (ICV) and stability on the ethnic populations data base

Relationship measure	ECV				ICV		Stability		Average rank	Overall rank
	% Concordance	Cramér	Corr. coef.	Overall average rank	Cophenetic corr.	Rank	Corr. coef.	Rank		
A Correlation coefficient	75·00	0·821	0·465	2·5	0·601	3	0·862	1	2·17	2
B Euclidean distance	75·00	0·824	0·462	2·5	0·655	2	0·617	3	2·50	3
C City-block distance	77·50	0·833	0·459	1	0·660	1	0·635	2	1·33	1

6.5 PERFORMANCE OF RELATIONSHIP MEASURES

The performances of the three relationship measures used in the present study, namely, correlation coefficient (A), Euclidean distance (B), and city-block distance (C), were assessed by averaging the performance of ordinal multidimensional scaling (2), single linkage (4), complete linkage (5) and *k*-means (7) using the coresponding relationship measures. These four taxonomic approaches used all three relationship measures. For example, for each evaluative criterion, the performance of the city-block distance (C) was assessed by averaging the performances of 2.C, 4.C, 5.C and 7.C.

Table 6.9 presents the performance and rankings of the three relationship measures on the ethnic populations data base according to external criterion validity, internal criterion validity and stability, and across all three evaluative criteria. The absolute performance values obtained for the relationship measures were quite similar to each other, especially according to the first two evaluative criteria, which suggests caution in giving weight to any differential ranking among them.

6.6 CLUSTER-BY-CLUSTER AGREEMENT BETWEEN EXPERT AND TAXONOMIC METHODS

The grouping produced by each of the 16 quantitative taxonomic methods on each ethnic population data set was cross-tabulated with the expert-based grouping of the ten ethnic populations into five pairs. All the resulting cross-classification tables were summed cell-by-cell in order to assess the rate of recovery of the native American, European, African, Oriental and Oceanic pairs of ethnic populations. There were 40 cross-classification tables: 20 for each data set, including two tables for each of the three ordinal multidimensional scaling methods and the Chernoff's faces method, as two judges produced separate clusterings for each of these four methods. Table 6.10 presents this cumulative information.

Table 6.11 was developed from Table 6.10 by computing the percentages of the expert-based groups, classified into the various clusters produced by the quantitative taxonomic methods. The European cluster had the highest recovery rate (91%). It was followed by the Oriental, Oceanic and African pairs at about the 70% recovery level. The American group had the lowest recovery rate (62%). The major forms of misclassification or confusion took place between the European and the African pairs (32%), and between the Oriental and the Oceanic pairs (44%). A moderate amount of misclassification (21%) was observed between the native American and the

Table 6.10

Cell-by-cell sum of cross-classification tables between groupings of the ethnic populations produced by the expert and each of the quantitative taxonomic methods

Expert-produced groups	Clusters produced by quantitative taxonomic methods					
	American	European	African	Oriental	Oceanic	
American	50	16	5	6	3	80
European	1	73	4	2	0	80
African	1	22	56	0	1	80
Oriental	1	4	1	58	16	80
Oceanic	0	3	1	19	57	80
	53	118	67	85	77	400

European pairs. Any other misclassification between ethnic population clusters had a level lower than 9%.

The highest cohesiveness inferred from the highest recovery rate observed between the English and Italian populations forming the European pair is in line with their being, from a geographical distance viewpoint, the two ethnic populations closest to each other. Again consistent with geographical distance was the lowest level of recovery rate obtained for the native American pair formed by Eskimos and Mayas.

Table 6.11

Percentages of each one of the expert-produced groups classified into the various clusters produced by the quantitative taxonomic methods on the ethnic populations data base

Expert-produced groups	Clusters produced by quantitative taxonomic methods					
	American	European	African	Oriental	Oceanic	
American	62·50	20·00	6·25	7·50	3·75	100·00
European	1·25	91·25	5·00	2·50	0·00	100·00
African	1·25	27·50	70·00	0·00	1·25	100·00
Oriental	1·25	5·00	1·25	72·50	20·00	100·00
Oceanic	0·00	3·75	1·25	23·75	71·25	100·00

The relatively large amount of misclassification obtained between the European and African pairs, and between the Oriental and Oceanic pairs is in line with their respectively tending to form two higher-order clusters, as was noted from the illustrative clustering processes described in § 6.3. This is again consistent with geographical distance distribution, as the European and African pairs and the Oriental and Oceanic pairs are respectively closer to each other than to any of the other three ethnic population pairs. These findings may also be explained by migration patterns (Piazza et al., 1975).

7

Comparative Performance of Quantitative Taxonomic Methods across Data Bases

7.1 INTRODUCTION

The performances of the 18 quantitative taxonomic methods, each defined by a particular combination of clustering approach and relationship measure, were studied first within each one of four real data bases and presented in Chapters 3 to 6, respectively. The rankings of the taxonomic methods as well as of the three relationship measures, according to various evaluative criteria and across them, are described in those chapters for each data base. An attempt was also made in each case to use the clustering results to clarify the relationships and differentiability among the various groups built by expert knowledge into the data.

In this chapter, the relative performance of the various quantitative taxonomic methods will be evaluated across data bases. The intention here is to elucidate more globally the rankings of the taxonomic methods and to explore their generalizability and variability across different fields.

The four data bases are examples of data corresponding to psychosocial, psychopathological, botanical and ethnic population fields. In turn, these fields represent illustrative examples of the kinds of information sources one finds in behavioral science (with the exception of the iris plant data, selected because of its frequent role as test data in clustering methodology).

7.2 COMPARATIVE PERFORMANCE OF THE 18 QUANTITATIVE TAXONOMIC METHODS

The first set of the results obtained across data bases corresponds to the comparison of the performance of the quantitative taxonomic methods according to each of the three major evaluative criteria (external criterion validity, internal criterion validity and replicability or stability) as well as across these three criteria. Then, a comparison of the ordinal multidimensional scaling methods and Chernoff's faces according to inter-rater reliability will be described. Finally, the performance of the three relationship measures used in this study will be compared across data bases.

Rankings of the quantitative taxonomic methods according to external criterion validity values on each of the four data bases and across them are presented in Table 7.1. For each data base, the percentages of the concordance with the field expert, the Cramér's statistic and the correlation coefficients between derived similarity matrices, were averaged across the half data sets and a ranking of the quantitative taxonomic methods was obtained according to each of these three measures of external criterion validity. Average ranks and overall ranks were then computed for each data base and are shown as such in Table 7.1. This table also shows both average ranks computed by averaging data base average ranks and final overall ranks.

The rankings on the treatment environments and the archetypal psychiatric patient data bases were highly correlated with each other (Spearman's rank $r = 0{\cdot}85$). By contrast, there was considerable variability in ranking between the first two data bases and the iris and the ethnic population data bases as well as between the last two (no rank correlation here was significantly different from zero). However, it was possible to detect some patterns in terms of those taxonomic methods clearly tending to rank well and others tending to rank poorly. The overall ranks indicate that the complete linkage methods (5.A, 5.C and 5.B) performed best. The Rubin-Friedman optimization techique (9.X) and the centroid linkage (6.A) also ranked well. At the poorest extreme of the rankings were the single linkage methods (4.A, 4.C and 4.B), next to these were NORMAP/NORMIX (10.X) and Chernoff's faces (3.X).

In appraising the above ranking of quantitative taxonomic methods, it should be kept in mind that in order to make the analysis commensurate, the cluster configuration produced by the taxonomic methods were required to have the same number of clusters as the expert's cluster configuration.

Table 7.2 exhibits the internal criterion validity values (cophenetic correlation coefficients) and rankings of the quantitative taxonomic methods on each of the four data bases and across them.

A comparison of the internal criterion validity rankings showed wide variability among data bases. No Spearman rank correlation coefficient computed between members of any pair of data bases turned out to be significantly different from zero.

However, on average, centroid linkage (6.A) performed best, followed by

Table 7.1

Ranking of quantitative taxonomic methods according to external criterion validity values (averaging % concordance, Cramér's statistics and correlation coefficients between derived similarity matrices) on each of four data bases (treatment environments, archetypal psychiatric patients, iris and ethnic populations) and across them

	Quantitative taxonomic methods	Treat. environ.		Arc. psych. pts.		Iris		Eth. popul.		Across fields	
		Avg. rank	Overall rank	Avg. rank	Overall rank	Avg. rank	Overall rank	Avg. rank	Overall rank	Avg. rank	Overall rank
1.A	Q-factor analysis, corr. coef.	8·8	9	10·7	10·5	14	14	10·2	9	10·91	13
2.A	Multidimen. scal., corr. coef.	10	10	10·7	10·5	10·7	10	11·2	12	10·63	12
2.B	Multidimen. scal., Euclid. dist.	12·7	13	11·7	12	8	8	5·2	5	9·47	10
2.C	Multidimen. scal., city-bl. dist.	14	14	13	13	9	9	3	4	9·75	11
3.X	Chernoff's faces	12	12	17	17	6·7	7	12·7	13	12·08	14
4.A	Single linkage, corr. coef.	16	16	15	15	15·8	16·5	10·5	10·5	14·33	18
4.B	Single linkage, Euclid. dist.	17·5	17·5	14·3	14	15·8	16·5	6·8	6·5	13·62	16
4.C	Single linkage, city-bl. dist.	17·5	17·5	15·7	16	15	15	6·8	6·5	13·75	17
5.A	Complete linkage, corr. coef.	1	1	5	5	2	2	2·3	2	2·58	1
5.B	Complete linkage, Euclid. dist.	6·8	7	9	9	3	3	2·3	2	5·28	3
5.C	Complete linkage, city-bl. dist.	6·7	6	5	5	4·2	4	2·3	2	4·55	2
6.A	Centroid linkage, corr. coef.	4·7	5	2	2	13·5	13	10·5	10·5	7·68	5
7.A	*k*-Means, corr. coef.	2	2	8	8	18	18	7·8	8	8·96	9
7.B	*k*-Means, Euclid. dist.	4	4	2	2	11	11	15	15	8·00	7
7.C	*k*-Means, city-bl. dist.	3·3	3	2	2	12·2	12	13·3	14	7·70	6
8.X	ISODATA	7·7	8	5	5	4·8	5	16	16	8·38	8
9.X	Rubin-Friedman	11·3	11	7	7	1	1			6·43	4
10.X	NORMAP/NORMIX	15	15	18	18	6·3	6			13·11	15

Table 7.2

Internal criterion validity values (cophenetic correlation coefficients) and rankings for the quantitative taxonomic methods on each of four data bases (treatment environments, archetypal psychiatric patients, iris and ethnic populations) and across them

Quantitative taxonomic methods		Treat. environ.		Arc. psych. pts		Iris		Eth. popul.		Across fields	
		Cophen. corr. coef.	Rank	Cophen. corr. coef.	Rank	Cophen. corr. coef.	Rank	Cophen. corr. coef.	Rank	Cophen. corr. coef.	Rank
1.A	Q-factor analysis, corr. coef.	0·3730	16	0·6230	16	0·7750	7	—0·1230	16	0·4120	17
2.A	Multidimen. scal., corr. coef.	0·5555	6	0·7405	3	0·8265	4	0·6530	9	0·6939	2
2.B	Multidimen. scal., Euclid. dist.	0·5678	4	0·7025	13	0·7655	9	0·6822	5	0·6795	3
2.C	Multidimen. scal., city-bl. dist.	0·4738	11	0·6920	14	0·7705	8	0·6825	3·5	0·6547	9
3.X	Chernoff's faces	0·3795	14	0·6112	17	0·5160	18	0·2620	14	0·4422	16
4.A	Single linkage, corr. coef.	0·2395	18	0·8340	1	0·8945	1	0·6805	6·5	0·6621	6
4.B	Single linkage, Euclid. dist.	0·3845	13	0·7445	2	0·8255	5	0·6925	1	0·6618	7
4.C	Single linkage, city-bl. dist.	0·4170	12	0·7285	4	0·8150	6	0·6860	2	0·6616	8
5.A	Complete linkage, corr. coef.	0·6115	2	0·7140	8·5	0·6185	14	0·6345	10	0·6446	10
5.B	Complete linkage, Euclid. dist.	0·5490	7	0·7155	6·5	0·7320	12	0·6720	8	0·6671	4
5.C	Complete linkage, city-bl. dist.	0·5350	8	0·7040	12	0·7290	13	0·6825	3·5	0·6626	5
6.A	Centroid linkage, corr. coef.	0·6720	1	0·6850	15	0·8895	2	0·6805	6·5	0·7318	1
7.A	*k*-Means, corr. coef.	0·5845	3	0·7155	6·5	0·8155	3	0·4370	13	0·6381	11
7.B	*k*-Means, Euclid. Dist.	0·5050	9	0·7270	5	0·7420	10	0·5710	12	0·6392	12
7.C	*k*-Means, city-bl. dist.	0·4990	10	0·7055	11	0·7360	11	0·5880	11	0·6321	13
8.X	ISODATA	0·5565	5	0·7140	8·5	0·6095	17	0·0875	15	0·4919	15
9.X	Rubin-Friedman	0·2900	17	0·7115	10	0·6110	16			0·5375	14
10.X	NORMAP/NORMIX	0·3745	15	0·0525	18	0·6125	15			0·3465	18

the ordinal multidimensional scaling methods using correlation coefficients and Euclidean distances (2.A and 2.B), the complete linkage methods using Euclidean and city-block distances (5.B and 5.C) and the single linkage methods (4.A, 4.B and 4.C). At the poorest extreme of the overall ranking were NORMAP/NORMIX (10.X), Q-factor analysis (1.A) and Chernoff's faces (3.X).

Table 7.3 presents the replicability correlation coefficients and rankings of the quantitative taxonomic methods on each of the four data bases and their averages across data bases.

There was noticeable variability in rankings across data bases. No rank correlation coefficient between these was significantly different from zero. However, in terms of overall ranks, the complete linkage methods using correlation coefficients and city-block distances (5.A and 5.C), ISODATA (8.X) and the Rubin-Friedman optimization technique (9.X) performed best. The poorest overall ranks were obtained for the single linkage method using Euclidean distances (4.B) (although this method ranked first in the treatment environments data base), NORMAP/NORMIX (10.X), ordinal multidimensional scaling using city-block and Euclidean distances (2.C and 2.B), and Chernoff's faces (3.X).

Table 7.4 shows rankings of the quantitative taxonomic methods computed by averaging ranks across the three major evaluative criteria (external criterion validity, internal criterion validity and replicability) on each of the four data bases (treatment environments, archetypal psychiatric patients, iris specimens, and ethnic populations), and an overall ranking computed by averaging ranks across data bases. Here, as in the case of external criterion validity, the rankings on the treatment environments and the archetypal psychiatric patients data bases were rather highly correlated (Spearman's rank $r = 0{\cdot}81$, significantly different from zero at $p < 0{\cdot}001$). Again, considerable ranking variability between the members of the other pairs of data bases was noted (none of these had a rank correlation coefficient significantly different from zero).

However, some taxonomic methods consistently ranked either very high or very low across data bases. The overall best ranked clustering approaches were complete linkage (5.A, 5.C and 5.B) and centroid linkage (6.A); and the poorest ranked were NORMAP/NORMIX (10.X) and Chernoff's faces (3.X).

The graphical representations of the entities obtained through ordinal multidimensional scaling and the Chernoff's faces method were inspected by human judges in order to produce a cluster configuration in each case. Table 7.5 shows rankings of these methods according to inter-rater reliability within and across data bases.

It can be seen that, overall, ordinal multidimensional scaling using

Table 7.3

Replicability correlation coefficients and rankings for the quantitative taxonomic methods on each of four data bases (treatment environments, archetypal psychiatric patients, iris and ethnic populations) and across them

Quantitative taxonomic methods		Treat. environ.		Arc. psych. pts		Iris		Eth. popul.		Across fields	
		Corr. coef.	Rank	Corr coef.	Rank	Corr. coef.	Rank	Corr. coef.	Rank	Corr. coef.	Rank
1.A	Q-factor analysis, corr. coef.	0·908	5	0·971	10	0·992	5	0·423	16	0·824	5
2.A	Multidimen. scal., corr. coef.	0·288	16	0·782	13	0·852	8	0·750	8	0·668	8
2.B	Multidimen. scal., Eculid. dist.	0·274	17	0·891	11	0·511	10	0·656	10	0·583	14
2.C	Multidimen. scal., city-bl. dist.	0·155	18	0·780	14	0·503	11	0·772	5·5	0·553	16
3.X	Chernoff's faces	0·358	15	0·438	17	0·973	7	0·528	13	0·574	15
4.A	Single linkage, corr. coef.	0·871	8	0·758	15	0·000	16·5	0·926	2·5	0·639	9
4.B	Single linkage, Euclid. dist.	1·000	1	0·068	18	0·000	16·5	0·540	11·5	0·402	18
4.C	Single linkage, city-bl. dist.	0·999	2	0·786	12	0·021	15	0·540	11·5	0·586	13
5.A	Complete linkage, corr. coef.	0·941	4	1·000	2	0·976	6	0·772	5·5	0·922	1
5.B	Complete linkage, Euclid. dist.	0·540	12	0·974	9	0·999	2·5	0·772	5·5	0·821	6
5.C	Complete linkage, city-bl. dist.	0·650	11	1·000	2	0·999	2·5	0·772	5·5	0·855	3
6.A	Centroid linkage, corr. coef.	0·968	3	0·998	5	0·042	14	0·926	2·5	0·734	7
7.A	*k*-Means, corr. coef.	0·889	7	0·995	8	—0·421	18	1·000	1	0·616	11
7.B	*k*-Means, Euclid. dist.	0·818	9	0·998	5	0·205	12	0·500	14	0·630	10
7.C	*k*-Means, city-bl. dist.	0·796	10	0·998	5	0·155	13	0·456	15	0·601	12
8.X	ISODATA	0·907	6	1·000	2	0·993	4	0·730	9	0·908	2
9.X	Rubin-Friedman	0·495	13	0·997	7	1·000	1			0·831	4
10.X	NORMAP/NORMIX	0·360	14	0·441	16	0·818	9			0·540	17

Table 7.4

Rankings of quantitative taxonomic methods computed by averaging ranks across three evaluative criteria (external criterion validity, internal criterion validity and replicability) on each of four data bases (treatment environments, archetypal psychiatric patients, iris and ethnic populations) and across data bases

	Quantitative taxonomic methods	Treat. environ.		Arc. psych. pts		Iris		Eth. popul.		Across fields	
		Avg. rank	Overall rank	Avg. rank	Overall rank	Avg. rank	Overall rank	Avg. rank	Overall rank	Avg. rank	Overall rank
1.A	Q-factor analysis, corr. coef.	10·00	9	12·17	15	8·67	6·5	13·67	15·5	11·13	16
2.A	Multidimen, scal., corr. coef.	10·67	12	8·83	10	7·00	4	9·67	11	9·04	8
2.B	Multidimen, scal., Euclid. dist.	11·33	13	12·00	14	9·00	8	6·67	8·5	9·75	10·5
2.C	Multidimen, scal., city-bl. dist.	14·33	17	13·67	16	9·33	9	4·33	2	10·42	14
3.X	Chernoff's faces	13·67	14·5	17·00	17	10·67	13	13·33	13	13·67	17
4.A	Single linkage, corr. coef.	14·00	16	10·33	11	11·33	14	6·5	6·5	10·54	15
4.B	Single linkage, Euclid. dist.	10·50	10·5	11·33	13	12·33	17	6·33	5	10·12	13
4.C	Single linkage, city-bl. dist.	10·50	10·5	10·67	12	12·00	15·5	6·67	8·5	9·96	12
5.A	Complete linkage, corr. coef.	2·33	1	5·17	2·5	7·33	5	5·83	4	5·17	1
5.B	Complete linkage, Euclid. dist.	8·67	8	8·17	9	5·83	1	5·17	3	6·96	4
5.C	Complete linkage, city-bl. dist.	8·33	7	6·33	5	6·50	3	3·67	1	6·21	2
6.A	Centroid linkage, corr. coef.	3·00	2	7·33	6	9·67	10	6·5	6·5	6·63	3
7.A	*k*-Means, corr. coef.	4·00	3	7·50	7	13·67	18	7·33	10	8·13	5
7.B	*k*-Means, Euclid. dist.	7·33	5	4·00	1	11·00	12	13·67	15·5	9·00	7
7.C	*k*-Means, city-bl. dist.	7·67	6	6·00	4	12·00	15·5	13·33	13	9·75	10·5
8.X	ISODATA	6·33	4	5·17	2·5	8·67	6·5	13·33	13	8·38	6
9.X	Rubin-Friedman	13·67	14·5	8·00	8	6·00	2			9·22	9
10.X	NORMAP/NORMIX	14·67	18	17·33	18	10·00	11			14·00	18

Table 7.5

Rankings of the ordinal multidimensional scaling and Chernoff's faces methods according to inter-rater agreement on each of four data bases (treatment environments, archetypal psychiatric patients, iris and ethnic populations) and then averaging ranks across data bases

Quantitative taxonomic methods		Treat. environ.		Arc. psych. pts		Iris		Eth. popul.		Across fields	
		Avg. rank	Overall rank	Avg. rank	Overall rank	Avg. rank	Overall rank	Avg. rank	Overall rank	Avg. rank	Overall rank
2.A	Multidimen. scal., corr. coef.	1·7	2	1	1	1·33	1	1·5	1·5	1·38	1
2.B	Multidimen. scal., Euclid. Dist.	3·0	3	2·33	2	4·0	4	3·0	3	3·08	3
2.C	Multidimen. scal., city-bl. dist.	1·3	1	2·66	3	3·0	3	1·5	1·5	2·12	2
3.X	Chernoff's faces	4·0	4	4	4	1·67	2	4·0	4	3·42	4

Table 7.6

Rankings of relationship measures computed by averaging ranks across three evaluative criteria (external criterion validity, internal criterion validity and replicability) on each of four data bases (treatment environments, archetypal psychiatric patients, iris and ethnic populations) and across data bases

Relationship measures	Treat. environ.		Arc. psych. pts		Iris		Eth. popul.		Across fields	
	Avg. rank	Overall rank	Avg. rank	Overall rank	Avg. rank	Overall rank	Avg. rank	Overall rank	Avg. rank	Overall rank
A Correlation coefficient	1·33	1	2·00	2	2·33	2·5	2·17	2	1·96	2
B Euclidean distance	1·67	2	2·17	3	1·33	1	2·50	3	1·92	1
C City-block distance	3·00	3	1·83	1	2·33	2·5	1·33	1	2·12	3

correlation coefficients, city-block and Euclidean distances ranked better than the Chernoff's faces method. The only exception was the case of the iris specimens data base, where only the ordinal multidimensional scaling method using correlation coefficients ranked better than Chernoff's faces.

7.3 COMPARATIVE PERFORMANCE OF THE RELATIONSHIP MEASURES

The comparative performance of the three main measures of relationship between entities used in the present study, namely, correlation coefficient, Euclidean distance and city-block distance was assessed by averaging the performance of ordinal multidimensional scaling, single linkage, complete linkage and *k*-means using the corresponding relationship measures. These four taxonomic approaches used all three relationship measures.

Table 7.6 presents a ranking of the relationship measures (correlation coefficient, Euclidean distance and city-block distance) computed by averaging ranks across all three general evaluative criteria on each of the four data bases and across them. The overall ranking was not conclusive since, on the various evaluative criteria and data bases, the absolute values for the relationship measures were quite similar to each other and their rankings were variable.

7.4 DISCUSSION AND SOME PRACTICAL RECOMMENDATIONS

Previous evaluative studies reported in the literature examined comparatively few clustering approaches, usually applied to only one data base and according to only one evaluative criterion. The aspects of the present study corresponding in design to those reports generally agreed with their findings.

Bartko, Strauss and Carpenter (1971) found that complete linkage produced clusterings of archetypal psychiatric patients closer to a classification established by experienced psychiatrists than did the Rubin-Friedman optimization technique. The same ordering of these two taxonomic methods according to external criterion validity was found in the present study, both on archetypal psychiatric patients and across data bases.

The internal criterion validity ranking of average centroid linkage, complete linkage and single linkage, in decreasing order of cophenetic correlation coefficients reported here, is in line with the findings of Sneath (1966), Cunningham and Ogilvie (1972), Baker (1974) and Hubert (1974). On the other hand, in contrast to the findings of the present study on each of the four data bases, Boyce (1969), when studying a set of hominoids, obtained higher cophenetic correlation coefficients for Q-principal components analysis than for average linkage.

The decreasing replicability values obtained by Rogers and Linden (1973) for Lorr and Radhakrishnan's (1967) factor analytic method, Ward's (1963) special agglomerative method (which, according to Everitt's 1974 Monte Carlo studies, behaves similarly to the average centroid linkage method), and Johnson's (1967) HICLUS program (assuming its single linkage option), agree with the replicability ranking across data bases found in the present study for Q-factor analysis, average centroid linkage and single linkage methods.

Proper consideration should be given to the variability in the rankings of quantitative taxonomic methods across evaluative criteria and data bases described in this study. However, in the light of persistent patterns noted here, it is in general reasonable to recommend the complete linkage as the preferred method, with centroid linkage as a close second option. If these procedures were not available to the user, then either the *k*-means method (which is particularly inexpensive in terms of computer time), or ISODATA or the Rubin-Friedman optimization technique (which is particularly expensive) may be selected.

Although ordinal multidimensional scaling did not rank high in this study, it may be a valuable approach in two regards. Firstly, for visualizing data and cluster structure, either as the final clustering procedure or preceding the use of another clustering methods. Visual inspection may be helpful in clarifying issues such as the number of clusters and particular cluster shapes (e.g. elongated clusters, which would suggest the use of the single linkage method). Secondly, it may be helpful for graphical representation and interpretation of groups found using other clustering methods.

The use of either multivariate normal mixture analysis (NORMAP/NORMIX) or facial representation of multidimensional points (Chernoff's faces) for clustering purposes, without previous successful testing on data of the type to be analyzed, does not seem to be warranted at this time.

No clear recommendation can be given regarding preferred relationship measures but in many cases correlation coefficients seem to do at least as well as Euclidean and city-block distances.

It should be noted that the results of a quantitative taxonomic study may be influenced not only by the cluster analytic method used (as shown in this study), but also by the scaling and transformation of the input data (as shown by Bartko, Strauss and Carpenter, 1971, among others).

The results of the present study, as indicated earlier, suggest preferred methods for conducting taxonomic studies in general as well as in fields corresponding to the data bases considered here. If a narrower or special problem, say a typology of heart disease patients, were under consideration, a set of pertinent patients with a known cluster structure could suggest the selection of a clustering procedure to be employed in an analysis.

8

Cluster Analysis of Quantitative Taxonomic Methods

8.1 INTRODUCTION

A number of classificatory arrangements of quantitative taxonomic methods has been proposed in the literature. The arrangements have typically been tentative and judgmental, rather than data-based. An illustrative classification is introduced and described in Chapter 2.

The possibility of increasing and refining our knowledge on the interrelationships among taxonomic methods by computing similarity measures (e.g. correlation coefficients) between them, and then cluster analyzing them, appears to be both intriguing and feasible in the context of the present study. Better information on the relations between taxonomic methods has obvious theoretical and methodological importance as well as potentially useful practical usage. For example, Anderberg (1973, p. 207) suggests that taxonomic methods included in the same cluster would be significantly redundant for a given situation while methods in different clusters might reveal different facets of the data set and might therefore profitably complement each other.

It is possible to derive a similarity matrix from a cluster configuration, as in the case of computing a cophenetic correlation coefficient (Sokal and Rohlf, 1962). If a number of taxonomic methods is employed on the same data set, an equal number of similarity matrices can be derived. Correlation coefficients computed between such similarity matrices can be used to estimate the relationships between those taxonomic methods and therefore to carry out a cluster analysis of them.

There seems to be no report in the literature on actual cluster analytic studies of quantitative taxonomic methods.

Table 8.1
Mean intercorrelations among quantitative taxonomic methods computed from the first half of the four data bases

	1.A	2.A	2.B	2.C	3.X	4.A	4.B	4.C	5.A	5.B	5.C	6.A	7.A	7.B	7.C	8.X	9.X	10.X
1.A	1·000	0·418	0·398	0·324	0·268	0·428	0·341	0·341	0·530	0·405	0·444	0·569	0·499	0·468	0·458	0·418	0·446	0·310
2.A	0·418	1·000	0·755	0·594	0·414	0·563	0·541	0·541	0·602	0·725	0·721	0·678	0·580	0·704	0·674	0·584	0·560	0·515
2.B	0·398	0·755	1·000	0·699	0·393	0·490	0·589	0·580	0·571	0·662	0·745	0·677	0·508	0·737	0·706	0·603	0·536	0·457
2.C	0·324	0·594	0·699	1·000	0·524	0·421	0·646	0·637	0·514	0·638	0·657	0·527	0·474	0·680	0·594	0·507	0·498	0·240
3.X	0·268	0·414	0·393	0·524	1·000	0·348	0·431	0·431	0·420	0·468	0·429	0·385	0·296	0·458	0·494	0·386	0·467	0·334
4.A	0·428	0·563	0·490	0·421	0·348	1·000	0·560	0·560	0·458	0·452	0·469	0·667	0·513	0·457	0·496	0·411	0·372	0·252
4.B	0·341	0·541	0·589	0·646	0·431	0·560	1·000	0·906	0·376	0·564	0·551	0·547	0·521	0·470	0·453	0·432	0·384	0·279
4.C	0·341	0·541	0·580	0·637	0·431	0·560	0·906	1·000	0·376	0·564	0·551	0·547	0·513	0·470	0·456	0·432	0·384	0·279
5.A	0·530	0·602	0·571	0·514	0·420	0·458	0·376	0·376	1·000	0·651	0·743	0·677	0·502	0·842	0·758	0·642	0·643	0·401
5.B	0·405	0·725	0·662	0·638	0·468	0·452	0·564	0·564	0·651	1·000	0·861	0·608	0·604	0·568	0·562	0·574	0·572	0·374
5.C	0·444	0·721	0·745	0·657	0·429	0·469	0·551	0·551	0·743	0·861	1·000	0·678	0·577	0·792	0·722	0·670	0·592	0·368
6.A	0·569	0·678	0·677	0·527	0·385	0·667	0·547	0·547	0·677	0·608	0·678	1·000	0·650	0·685	0·662	0·611	0·541	0·412
7.A	0·499	0·580	0·508	0·474	0·296	0·513	0·521	0·513	0·502	0·604	0·577	0·650	1·000	0·596	0·595	0·464	0·420	0·242
7.B	0·468	0·704	0·737	0·680	0·458	0·457	0·470	0·470	0·842	0·568	0·792	0·685	0·596	1·000	0·867	0·571	0·623	0·402
7.C	0·458	0·674	0·706	0·594	0·494	0·496	0·453	0·456	0·758	0·562	0·722	0·662	0·595	0·867	1·000	0·547	0·628	0·389
8.X	0·418	0·584	0·603	0·507	0·386	0·411	0·432	0·432	0·642	0·574	0·670	0·611	0·464	0·571	0·547	1·000	0·573	0·412
9.X	0·446	0·560	0·536	0·498	0·467	0·372	0·384	0·384	0·643	0·572	0·592	0·541	0·420	0·623	0·628	0·573	1·000	0·356
10.X	0·310	0·515	0·437	0·240	0·334	0·252	0·279	0·279	0·401	0·374	0·368	0·412	0·242	0·402	0·389	0·412	0·356	1·000

Table 8.2

Mean intercorrelations among quantitative taxonomic methods computed from the second half of the four data bases

	1.A	2.A	2.B	2.C	3.X	4.A	4.B	4.C	5.A	5.B	5.C	6.A	7.A	7.B	7.C	8.X	9.X	10.X
1.A	1·000	0·590	0·422	0·371	0·314	0·435	0·352	0·379	0·564	0·462	0·419	0·607	0·598	0·505	0·520	0·441	0·540	0·231
2.A	0·590	1·000	0·537	0·528	0·339	0·761	0·468	0·482	0·627	0·612	0·545	0·816	0·640	0·645	0·655	0·493	0·496	0·291
2.B	0·422	0·537	1·000	0·790	0·459	0·497	0·561	0·565	0·711	0·838	0·781	0·625	0·529	0·550	0·564	0·659	0·833	0·203
2.C	0·371	0·528	0·790	1·000	0·403	0·449	0·548	0·574	0·593	0·765	0·787	0·555	0·420	0·493	0·488	0·554	0·604	0·179
3.X	0·314	0·339	0·459	0·403	1·000	0·347	0·468	0·418	0·428	0·493	0·384	0·335	0·312	0·394	0·392	0·410	0·498	0·162
4.A	0·435	0·761	0·497	0·449	0·347	1·000	0·531	0·563	0·505	0·472	0·455	0·693	0·522	0·517	0·519	0·356	0·451	0·243
4.B	0·352	0·468	0·561	0·548	0·468	0·531	1·000	0·837	0·446	0·536	0·537	0·476	0·426	0·573	0·581	0·400	0·506	0·227
4.C	0·379	0·482	0·565	0·574	0·418	0·563	0·837	1·000	0·431	0·527	0·561	0·509	0·447	0·553	0·560	0·390	0·488	0·228
5.A	0·564	0·627	0·711	0·593	0·428	0·505	0·446	0·431	1·000	0·690	0·624	0·714	0·610	0·576	0·575	0·533	0·745	0·239
5.B	0·462	0·612	0·838	0·765	0·493	0·472	0·536	0·527	0·690	1·000	0·839	0·651	0·527	0·620	0·589	0·718	0·654	0·211
5.C	0·419	0·545	0·781	0·787	0·384	0·455	0·537	0·561	0·624	0·839	1·000	0·612	0·462	0·546	0·550	0·685	0·634	0·237
6.A	0·607	0·816	0·625	0·555	0·335	0·693	0·476	0·509	0·714	0·651	0·612	1·000	0·730	0·657	0·668	0·566	0·605	0·293
7.A	0·598	0·640	0·529	0·420	0·312	0·522	0·426	0·447	0·610	0·527	0·462	0·730	1·000	0·634	0·642	0·513	0·610	0·216
7.B	0·505	0·645	0·550	0·493	0·394	0·517	0·573	0·553	0·576	0·620	0·546	0·657	0·634	1·000	0·950	0·532	0·579	0·212
7.C	0·520	0·655	0·564	0·488	0·392	0·519	0·581	0·560	0·575	0·589	0·550	0·668	0·642	0·950	1·000	0·560	0·621	0·214
8.X	0·441	0·493	0·659	0·554	0·410	0·356	0·400	0·390	0·533	0·718	0·685	0·566	0·513	0·532	0·560	1·000	0·709	0·234
9.X	0·540	0·496	0·833	0·604	0·498	0·451	0·506	0·488	0·745	0·654	0·634	0·605	0·610	0·579	0·621	0·709	1·000	0·197
10.X	0·231	0·291	0·203	0·179	0·162	0·243	0·227	0·228	0·239	0·211	0·237	0·293	0·216	0·212	0·214	0·234	0·197	1·000

8.2 CORRELATIONS AMONG QUANTITATIVE TAXONOMIC METHODS

The 18 quantitative taxonomic methods studied here, each defined as a combination of a clustering approach and a relationship measure, are described in detail in § 2.2. A rather lengthy procedure was used to compute the intercorrelations between these taxonomic methods on each of eight data sets, two from each of the four data bases (treatment environments, archetypal psychiatric patients, iris specimens and ethnic populations) studied in Chapters 3 to 6.

First, a matrix of similarities between entities was derived from the clustering output produced by each taxonomic method in a given data set. The entries of each similarity matrix were determined to be 1 if the corresponding entities belonged to the same cluster, and 0 otherwise.

Then, the relationship between two taxonomic methods was estimated by computing a product-moment correlation coefficient between corresponding entries of the similarity matrices derived from their outputs on the same data set. In this way, a correlation matrix among the 18 quantitative taxonomic methods was computed for each of the eight data sets.

The resulting eight correlation matrices were divided into two groups: one including the matrices for the data sets corresponding to the first half of the four data bases, and the other including those for the data sets corresponding to the second half. Thus, each group of data sets and matrices represented all four fields of application (each two data sets being the result of randomly halving an original data base). Then, a matrix of mean intercorrelations among quantitative taxonomic methods was computed for each group by averaging the corresponding entries of the four correlation matrices included in that group.

The matrix of mean intercorrelations among taxonomic methods for the first-half data sets is presented in Table 8.1 and the matrix for the second-half data sets is presented in Table 8.2.

These correlation matrices were used for carrying out cluster anlyses of the quantitative taxonomic methods (§ 8.3) and, indirectly, for assessing the interrelations among three relationship measures (§ 8.4).

8.3 CLUSTER ANALYSIS OF THE QUANTITATIVE TAXONOMIC METHODS

The complete linkage agglomerative hierarchical method (Johnson, 1967), using correlation coefficients, was selected as the main procedure for carrying out the cluster analysis of quantitative taxonomic methods. This method was selected because it had the best ranking across all general evaluative criteria and across all four data bases, as shown in Chapter 7. The second method used for cluster analyzing the quantitative taxonomic methods was ordinal multidimensional scaling (Kruskal, 1964a, b), using correlation coefficients. This method was chosen to provide a graphical, visualizable representation of the relationships among the quantitative taxonomic methods, which would complement the numerical complete linkage clustering approach. Both the complete linkage and the ordinal multidimensional scaling methods are described in § 2.2.

The two matrices of intercorrelations among the 18 quantitative taxonomic methods were cluster anlayzed first, using the complete linkage agglomerative method. The dendrograms of 8.1 and 8.2 represent the two hierarchical cluster configurations produced, respectively, from the matrix corresponding to the first-half of the four data bases (Table 8.1) and from the matrix corresponding to the second-half of the four data bases (Table 8.2). In each case, the whole clustering process is displayed from the

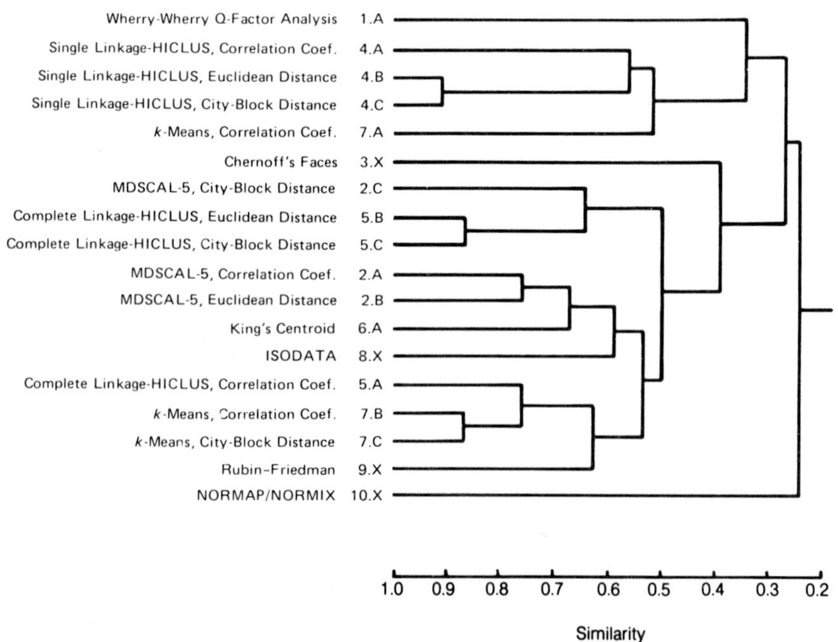

Fig. 8.1 Dendrogram representing the complete linkage cluster analysis of the 18 quantitative taxonomic methods using mean intercorrelations among methods computed from the first half of the four data bases.

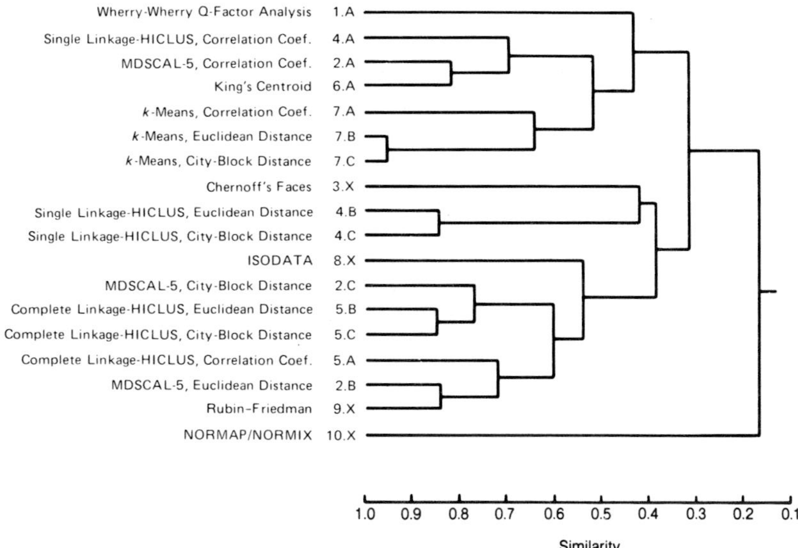

Fig. 8.2 Dendrogram representing the complete linkage cluster analysis of the 18 quantitative taxonomic methods using mean intercorrelations among methods computed from the second half of the four data bases.

beginning on the left where each taxonomic method is a single cluster, to the end where all taxonomic methods are lumped into one cluster. Looking at both dendrograms, it is striking that in both of them NORMAP/NORMIX (10.X), Q-factor analysis (1.A) and Chernoff's faces (3.X) seem to be quite independent of all other methods and of each other, as they do not cluster with other methods until quite late in the agglomerative process. It can also be seen that in both dendrograms the three first pairs formed are 4.B-4.C, 5.B-5.C and 7.B-7.C, which suggests that both clustering technique (i.e. 4, single linkage; 5, complete linkage; and 7, *k*-means) and similarity between the two distance functions (B, Euclidean distance; and C, city-block distance) strongly influenced the clustering process. Additional observations about the interrelations among the relationship measures, made from these figures and those to follow, are formulated in § 8.4.

The matrices of mean intercorrelations among quantitative taxonomic methods presented in Tables 8.1 and 8.2 were next analyzed through ordinal multidimensional scaling, using correlation coefficients, in order to have a visualizable representation of the configuration of quantitative taxonomic methods. The two-dimensional representation of taxonomic methods obtained from Table 8.1 is shown in Fig. 8.3, and that from Table 8.2 is

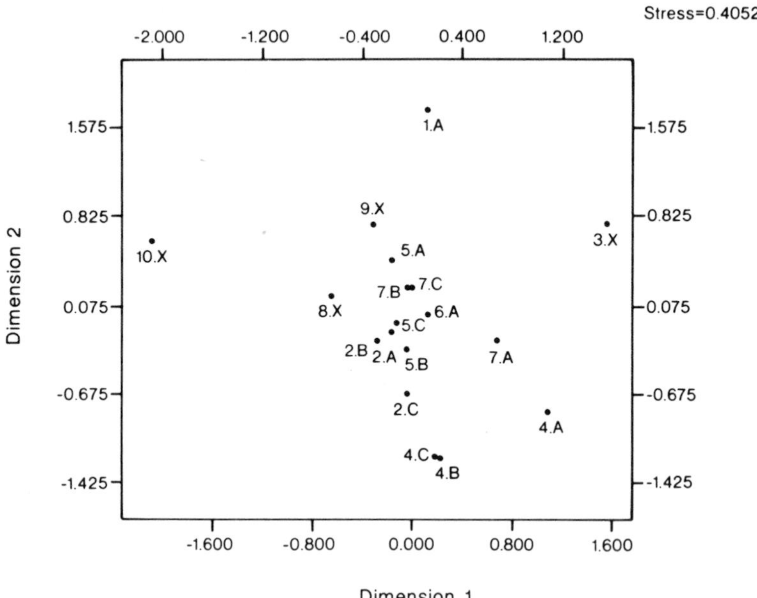

Fig. 8.3 Ordinal multidimensional scaling representation of the 18 quantitative taxonomic methods using mean intercorrelations among methods computed from the first half of the four data bases.

given in Fig. 8.4. The latter figure, with a stress value of 0·007, shows only that NORMAP/NORMIX (10.X) is greatly separated from the other 17 methods, all of which appear almost concentrated in one point. Figure 8.3 (stress value 0·405) has more discriminating information about the various taxonomic methods. It shows that, again, NORMAP/NORMIX (10.X) is quite separated from the other methods but also that Q-factor analysis (1.A) and Chernoff's faces (3.X) are considerably separated from the others as well as from each other. The single linkage methods (4.A, 4.B and 4.C) are located in the lower right region of the graph and somewhat separated from the other methods. Thus, the ordinal multidimensional scaling configuration of the 18 taxonomic methods presented in Fig. 8.3 appears to be quite consistent with the major features of the complete linkage hierarchical configurations shown in Fig. 8.1 and 8.2.

Attention was then focused on studying the cluster configuration of the ten basic clustering techniques, namely, Q-factor analysis (1), ordinal multidimensional scaling (2), Chernoff's faces (3), single linkage (4), complete linkage (5), centroid linkage (6), *k*-means (7), ISODATA (8), the Rubin-Friedeman optimization technique (9) and NORMAP/NORMIX (10).

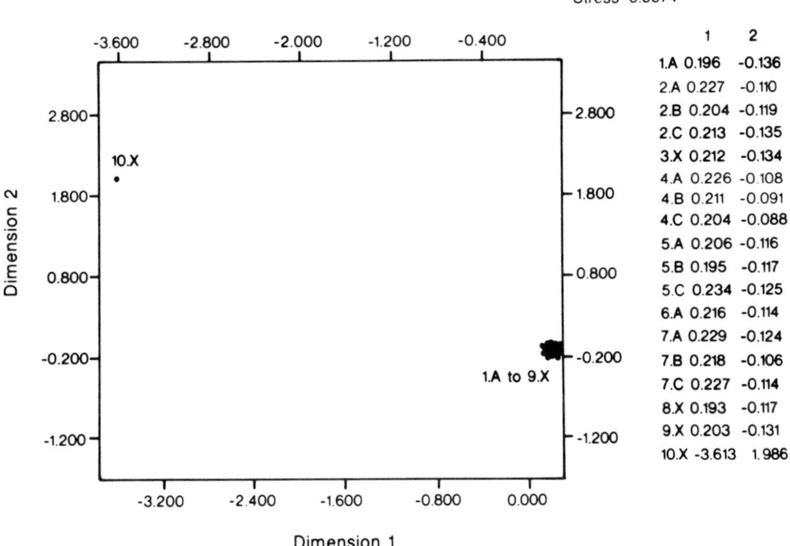

Fig. 8.4 Ordinal multidimensional scaling representation of the 18 quantitative taxonomic methods using mean intercorrelations among methods computed from the second half of the four data bases.

Their configuration was explored by cluster analysing exclusively the ten quantitative taxonomic methods which do not use relationship measures B (Euclidean distance) or C (city-block distance). This left a set of quantitative taxonomic methods (1.A, 2.A, 3.X, 4.A, 5.A, 6.A, 7.A, 8.X, 9.X and 10.X), which included one representative of each of the ten basic clustering approaches, each using either correlation coefficients (A) or none of our relationship measures in particular (X).

First, two matrices of mean intercorrelations among these ten methods were respectively obtained for the first half of the four data bases and for the second half of the data bases, by eliminating the non-pertinent columns and rows from the matrices presented in Tables 8.1 and 8.2. Then, each one of these correlation matrices was cluster analyzed, first through the complete linkage method and then through ordinal multidimensional scaling.

Figures 8.5 and 8.6 exhibit dendrograms representing the two hierarchical cluster configurations produced, respectively, from the matrix corresponding to the first half of the data bases and from the matrix corresponding to the second half of the data bases. It can be seen, in both dendrograms, that ordinal multidimensional scaling (2.A) and centroid linkage (6.A) were

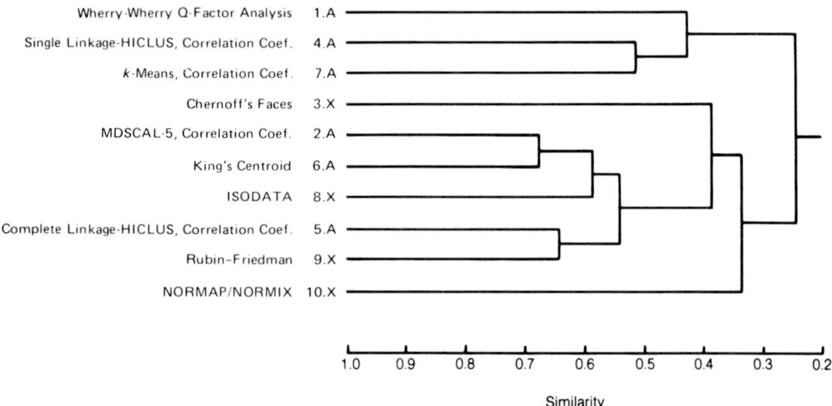

Fig. 8.5 Dendrogram representing the complete linkage cluster analysis of the ten basic clustering methods using mean intercorrelations among methods computed from the first half of the four data bases.

the first two taxonomic methods to cluster, complete linkage (5.A) and the Rubin-Friedman optimization technique (9.X) clustered next, and then all these four methods with the addition of ISODATA (8.X) formed a large cluster at a similarity level of about 0·55. Chernoff's faces (3.X) and NORMAP/NORMIX (10.X) appeared conspicuously isolated from each other and from the other methods as they were integrated very late in the agglomerative process. The cluster configuration presented in Fig. 8.5 is entirely overlapping with the configuration for the corresponding ten

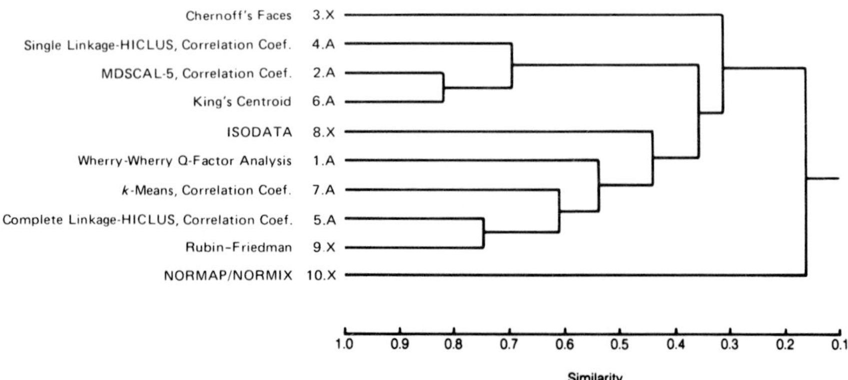

Fig. 8.6 Dendrogram representing the complete linkage cluster analysis of the ten basic clustering methods using mean intercorrelations among methods computed from the second half of the four data bases.

methods which can be abstracted from the dendrogram presented in Fig. 8.1. The cluster configuration presented in Fig. 8.6 shows some differences from its sister configuration in Fig. 8.5, in that Q-factor analysis (1.A) appeared to be somewhat less isolated from the other methods and that single linkage (4.A) and *k*-means (7.A) did not form a basic cluster (although it should be noted that it was rather late in the agglomerative process when these two methods clustered in Fig. 8.5).

Next, the matrices of mean intercorrelations among the ten basic clustering methods were analyzed through ordinal multidimensional scaling. The two-dimensional representation of the ten taxonomic methods obtained from the correlation matrix shown in Table 8.1 is presented in Fig. 8.7 and that obtained from the correlation matrix shown in Table 8.2 is presented in Fig. 8.8. The representation in Fig. 8.7 had a stress value of 0·307. It shows that Q-factor analysis (1.A), Chernoff's faces (3.X) and NORMAP/ NORMIX (10.X) are quite distant from each other and from the other methods, that ordinal multidimensional scaling (2.A), complete linkage (5.A), centroid linkage (6.A), ISODATA (8.X) and the Rubin-Friedman

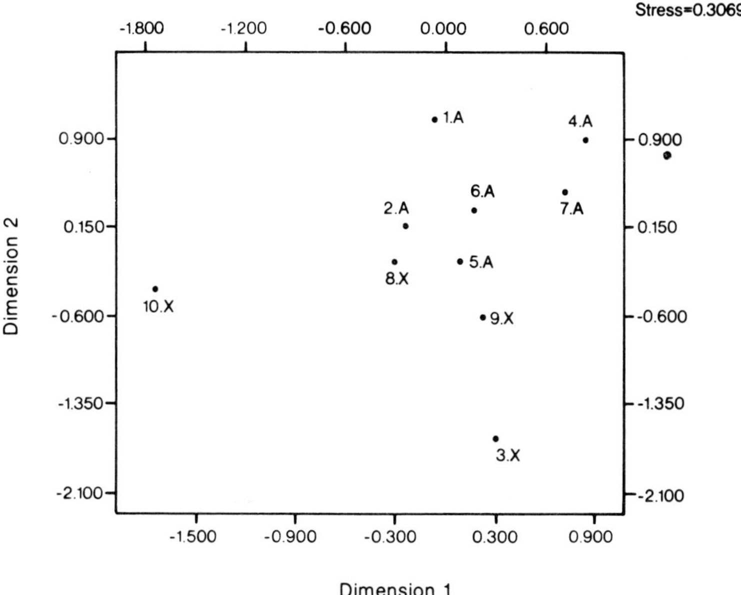

Fig. 8.7 Ordinal multidimensional scaling representation of the ten basic clustering methods using mean intercorrelations among methods computed from the first half of the four data bases.

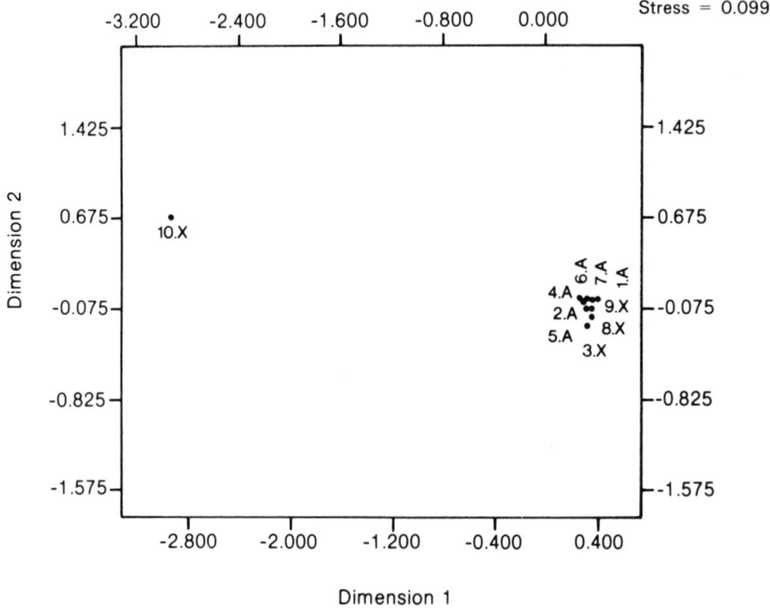

Fig. 8.8 Ordinal multidimensional scaling representation of the ten basic clustering methods using mean intercorrelations among methods computed from the second half of the four data bases.

optimization technique (9.X) seem to form a cluster, and that single linkage (4.A) and, to some extent, *k*-means (7.A) were located marginally on the graph somewhat separated from the other methods, and relatively close to each other. All of this is quite consistent with the complete linkage cluster configuration presented in Fig. 4.5. Figure 4.8, which has a stress of 0·010, essentially shows that NORMAP/NORMIX (10.X) is greatly separated from the other nine methods and that these, in view of that contrast, appear very close to each other.

From the cluster configurations of quantitative taxonomic methods presented in Fig. 8.1 to 8.8, the following conclusions can be drawn. Q-factor analysis, Chernoff's faces and, particularly, pattern clustering by multivariate mixture analysis (NORMAP/NORMIX) appeared to be quite different and isolated both from each other and from the other methods. Focusing attention on the basic clustering methods by making relationship measures uniform, it appeared that complete linkage and the Rubin-Friedman optimization method always clustered together and frequently formed a large cluster with ISODATA as well as with ordinal mutlidimensional scaling and centroid linkage (a pair which also always clustered together). The *k*-means method sometimes joined the large cluster and at other times formed a cluster with the single linkage method which, in general, tended to be separated from the other taxonomic methods.

The isolated positions found for Q-factor analysis, Chernoff's faces and NORMAP/NORMIX are not very surprising as each of them, especially the last two, represent quite particular approaches to quantitative taxonomy as described in Chapter 2. As shown comparatively in § 7.2, these three taxonomic methods turned out to be those with the poorest average performance across all evaluative criteria and all four data bases used.

The larger cluster included five basic clustering methods (complete linkage, average linkage, ISODATA, ordinal multidimensional scaling and the Rubin-Friedman optimization method) which were among the six methods with the highest performance across evaluative criteria and data bases. The third highest performance was attained by the *k*-means method which on one occasion joined this cluster.

The rather frequent separation of the *k*-means method (a nearest-centroid sorting method with a fixed number of clusters) from ISODATA (a nearest-centroid sorting method with a flexible number of clusters) seems to suggest a qualitatively differentiating influence on clustering output of the sophisticated devices (e.g. splitting and lumping) built into ISODATA.

The single linkage method was consistently separated from the large cluster of taxonomic methods, which is not surprising as one of the few persistent findings and comments in the literature comparing taxonomic methods refers to the particular cluster shapes (i.e. serpentine, elongated) for which this method seems to be suitable (Everitt, 1974, p. 84). Also in line with these results are Sneath's (1966) findings, using a group of randomly spaced points, which indicate that the correlation between the two similarity matrices, respectively derived from average linkage and complete linkage configurations, was higher than the correlations between either of them and the similarity matrix derived from a single linkage configuration.

8.4 INTERRELATIONS AMONG THE RELATIONSHIP MEASURES

Although the main emphasis of the study described in this chapter has been on the cluster analysis of the full quantitative taxonomic methods, it was possible also to obtain some information about interrelations among the three relationship measures used in the study, namely, correlation coefficient, Euclidean distance and city-block distance.

The first comparison was addressed to the matrices of intercorrelations among quantitative taxonomic methods separately computed from the first half and the second half of the four data bases and presented in Tables 8.1 and 8.2, respectively. The specific correlation coefficients examined were those corresponding to each of the three forms (defined by relationship measure employed) of clustering methods 2 (ordinal multidimensional scaling), 4 (single linkage), 5 (complete linkage) and 7 (*k*-means). These four clustering methods were those used in connection with all three relationship measures.

It was seen in both correlation matrices that the correlation between the forms of a given taxonomic method using Euclidean distance and city-block distance was consistently higher than the correlations of the form using the correlation coefficient with either of the forms using Euclidean distance or city-block distance. The only exception was that in Table 8.1 the correlation between the forms of ordinal multidimensional scaling, using the correlation coefficient and the Euclidean distance, was higher than the correlation between the forms using Euclidean and city-block distances which, however, was higher than the remaining correlation. The mean correlation coefficients between each two of the three forms of the four relevant clustering methods, computed by averaging the appropriate correlation coefficients in Tables 8.1 and 8.2, were as follows:

$$\bar{r}_{\mathrm{A\cdot B}} = 0{\cdot}620 \qquad \bar{r}_{\mathrm{A\cdot C}} = 0{\cdot}606 \qquad \bar{r}_{\mathrm{B\cdot C}} = 0{\cdot}844\,.$$

The second examination was addressed to both dendrograms depicting the complete linkage agglomerative cluster analyses of the 18 quantitative taxonomic methods (Fig. 8.1 and 8.2). Such inspection indicated that for each of the four methods involving correlation coefficients, Euclidean distances and city-block distances, the forms using Euclidean and city-block distances consistently clustered with each other much earlier than with the form using correlation coefficients. The only exception, again, was ordinal multidimensional scaling using Euclidean distances (2.B), which in Fig. 8.1 (but not in Fig. 8.2) clustered earlier with the form using correlation coefficients (2.A) than with that using city-block distances (2.C).

The last examination of the interrelations among the three relationship measures was conducted on the ordinal multidimensional scaling representation of the 18 quantitative taxonomic methods shown in Fig. 8.3 (the other representation shown in Fig. 8.4 did not provide much discriminating information among those 17 taxonomic methods which were greatly separated from NORMAP/NORMIX). That examination indicated that for each of the four methods using correlation coefficients, Euclidean distances and city-block distances, those combinations using Euclidean and city-block distances appeared to be closer to each other than to that using correlation coefficients. Again, as in previous cases, the only exception was that the ordinal multidimensional scaling method using Euclidean distances (2.B) appeared closer to that using correlation coefficients (2.A) than to that using city-block distances (2.C).

The strong tendency showed by quantitative taxonomic methods (representing a given clustering approach) using Euclidean distances and city-block distances to correlate higher and then cluster with each other rather than with the corresponding method using correlation coefficients, is in line with the study of interrelations among relationship measures carried out by Green and Rao (1969). They found the following intercorrelation values: 0·96 between squared Euclidean distance and squared city-block distance; 0·21 between squared Euclidean distance and correlation coefficient, and 0·24 between squared city-block distance and correlation coefficient. These authors also found that this pattern of intercorrelations was clearly noticeable on an ordinal multidimensional scaling representation of several relationship measures.

8.5 CLUSTER ANALYSIS OF EIGHT DATA SETS REPRESENTING DATA FROM FOUR DIFFERENT DISCIPLINARY FIELDS

Elaborating on the cluster analyses of quantitative taxonomic methods, a cluster analytic study of the eight data sets obtained by randomly halving each of the four original data bases (treatment environments, archetypal psychiatric patients, iris specimens and ethnic populations) was carried out, using the similarities among cluster configurations obtained on those data sets.

The initial step in this study was the computation of a matrix of intercorrelations among the eight data sets (Table 8.3), where each data set was represented by a matrix of intercorrelations among the similarity matrices derived from the cluster configurations produced by the quantitative taxonomic methods on that data set.

The correlation matrix presented in Table 8.3 was then cluster analyzed by the complete linkage method. Figure 8.9 exhibits a dendrogram representing the obtained hierarchical cluster configuration of data sets. It can be clearly seen that the members of the pair of data sets produced by randomly halving each original data base, clustered first with each other, i.e. 1-2 (treatment environments measured in terms of psychosocial climate), 3-4 (archetypal psychiatric patients measured on psychopathological symptoms), 5-6 (iris plants measured along morphological variables) and 7-8 (ethnic populations assessed in terms of genetic biochemical variables).

Table 8.3
Intercorrelations among the eight data sets obtained by randomly halving the four data bases. Each data set was represented by a matrix containing intercorrelations among the quantitative taxonomic methods on that data set

	1	2	3	4	5	6	7	8
1	1·000	0·546	0·323	0·176	—0·024	—0·067	0·010	0·173
2	0·546	1·000	0·538	0·397	—0·026	—0·042	—0·177	—0·094
3	0·323	0·538	1·000	0·913	0·039	0·205	0·010	0·036
4	0·176	0·397	0·913	1·000	0·012	0·275	0·148	0·129
5	—0·024	—0·026	0·039	0·012	1·000	0·371	0·259	0·264
6	—0·067	—0·042	0·205	0·275	0·371	1·000	0·322	0·417
7	0·010	—0·177	0·010	0·148	0·259	0·322	1·000	0·677
8	0·173	—0·094	0·036	0·129	0·264	0·417	0·677	1·000

Then, two higher-order clusters were neatly formed: one including the treatment environments (psychosocial) and the archetypal psychiatric patients (psychopathological) data bases, and the other including the iris (botanical) and the ethnic populations (genetic) data bases. The former cluster includes what is usually considered behavioral data and the latter, biological data.

Next, the eight data sets were anlyzed through ordinal multidimensional scaling using the correlation matrix presented in Table 8.3. The obtained two-dimensional representation of the data sets, which had a stress value of 0·143, is shown in Fig. 8.10. The configuration is entirely consistent with the configuration produced by the complete linkage method: data sets 1 and 2, 3 and 4, 5 and 6, and 7 and 8, closely clustered with each other first. Then, the behavioral data sets appeared to cluster in the upper right corner and the biological data sets in the lower left corner of the field.

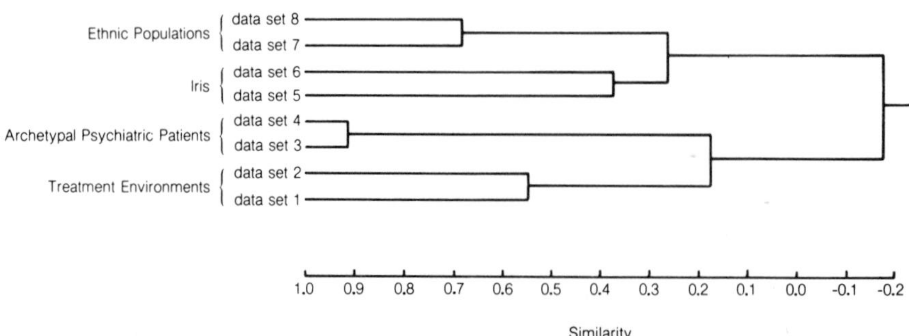

Fig. 8.9 Dendrogram representing the complete linkage cluster analysis of the 8 data sets obtained by randomly halving four data bases.

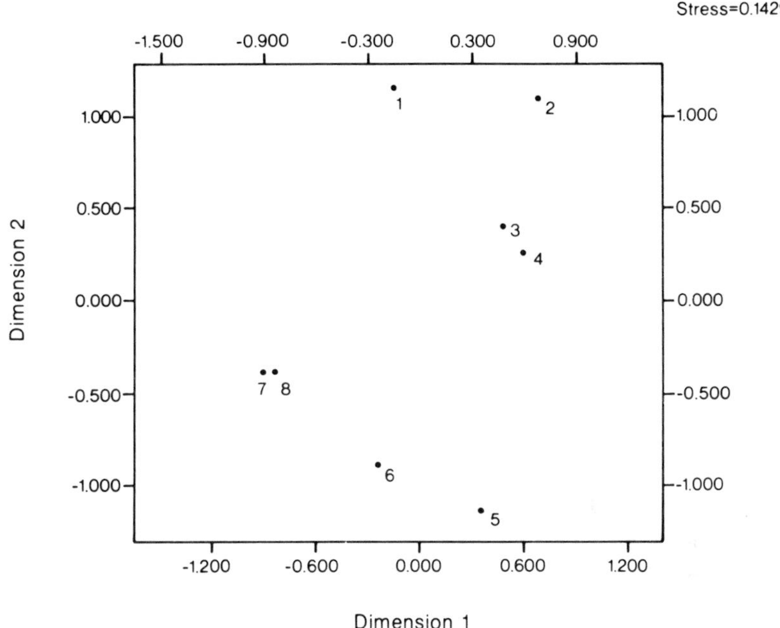

Fig. 8.10 Ordinal multidimensional scaling representation of the 8 data sets obtained by randomly halving four data bases.

The structure and hierarchy among data sets elucidated in this short study illustrate the promising contribution of quantitative taxonomy to the understanding of the interrelations of different kinds of information. This may be very valuable for designing and implementing biopsychosocial and, in general, interdisciplinary approaches in behavioral science research.

9

Overview and Uses

The main emphasis of the exposition until now has been on the empirical evaluation of the major clustering algorithms proposed in the literature. Associated with this is some necessary discussion of the representation of multivariate data since these can serve as a precursor to achieving clusters. Conceptually, such procedures do not serve as clustering algorithms and too often they are confused with grouping methods. A prime example of this is in the use of factor analysis as a clustering rule when it is essentially a data representation technique that reduces the dimensionality of multiple measurements from which clustering can then take place. More could have been said about graphical representation of multivariate data vectors. Some ideas and evaluations along these lines appear in Mezzich and Worthington (1978).

To satisfy their intellectual curiosity and to provide benchmarks for other investigators, the authors mounted a Herculean effort to try almost as many clustering techniques as have been proposed, each with varying similarity coefficients, and have applied them to a number of data bases. Chapters 3, 4, 5 and 6 bear witness to this prodigious task. It was thought essential also to vary the content of the data bases along some spectrum. This has been accomplished by providing physiological data bases (iris and ethnic populations) and psychological data bases (archetypal psychiatric patients and treatment environments). A strong reason for employing the iris data is that they serve as an anchor data base for many in the clustering field.

We have already remarked that data bases do not appear toc frequently in the open literature and that one is either a captive of what has appeared or must develop one's own data, the data bases given in each of Chapters 3, 4, 5 and 6 resulted from these constraints. Yet they are full-bodied and varied enough to demonstrate how well or how badly clustering algorithms perform.

The major results to guide any investigator who is embarking on clustering multivariate data vectors appear in Chapters 7 and 8. There, one can see not only how well each clustering algorithm performs on varying data bases but also how the techniques themselves group together. Taxonomic methods that cluster should, essentially, give similar results for a given data base thus permitting a selection based on other factors such as existing computer software or computational ease. A taxonomic procedure situated in another cluster and applied in conjunction could possibly complement the first procedure in that it could reveal differing facets of the data. Thus, if two taxonomic procedures are to be employed it may be expeditious in some cases to use two disparate ones, that is, two that do not cluster, and our results indicate which clustering procedures are far apart for this purpose.

It should be instructive to report on how the results of this monograph can be applied. One of the authors was faced with analyzing a large data base subsequent to the completion of the results given in the previous chapters. At hand was a sample of approximately 900 urban families, taken from 1960/1961 US census data, for each of whom the expenditure patterns of households across 12 categories (e.g. food, rent, clothing, education) were given. Each household, therefore, was represented by a 12-dimensional vector whose coordinates were the proportions of dollars of total income spent in each category.

It is of interest to learn whether the household expenditure patterns group in any meaningful way. Micro-economists would suggest that each household is governed by its individual utility function and this suggests no real groupings. Institutional economists would feel happier if groupings occurred according to such variables as income, socio-economic status, size of family, etc. Moreover, if such groupings occurred, there would be interest in the homogeneity found in each cluster. We have addressed ourselves to that question in what has been discussed in our clustering inquiries in previous chapters and have applied levels of homogeneity to determine cluster configurations.

We now place ourselves in the role of the investigator who has extracted information given in the previous chapters, has some computer time and clustering software available, and is eager to find clusters existing in a data base and then to attempt measurement of degrees of homogeneity in clusters arising for their particular data base.

The first thing that will strike the household expenditure investigator is the size of the data base since it will lead to a similarity matrix of a very large order. Each element in the matrix will be the distance in a 12-dimensional space (actually 11 since the 12 expenditure proportions across each household vector sum to one) between two household expenditure patterns. In the anlysis that was attempted, Mahalanobis distances between all possible pairs of 896 household expenditures were computed (four of the 900 fell by the wayside) to form the elements of the similarity matrix. The covariance matrix employed was that for the total data universe. It was not feasible to recompute the covariance matrices for each clustering configuration that was produced by a clustering algorithm in order to compute the Mahalanobis distances. There is an assumption here that the Mahalanobis distances were not too affected by this attempt to provide computational reasonableness.

The use of any clustering algorithm on a large similarity matrix (896 × 896) requires serious planning. One of the simplest hierarchical algorithms to employ is the centroid linkage procedure and our previous results suggest that it does well. It is one of those with the best performance across evaluative criteria and data bases. However, the size of the data base on household expenditure patterns is far and away much larger than the four data bases that form the heart of the study in previous chapters.

The King clustering program that provides centroid linkage hierarchical clustering (in this case from 896 clusters of one element to one cluster of 896 elements) would require modification, yet it seemed the best from the point of view of costs of computer time and costs of modification of existing software.

This clustering procedure was applied and careful examination was given to the resulting clusters as they were spawned by centroid linkage. Naturally, for each household, vital statistics such as family income, family size, age of head of family and occupation of family head were carried along in the computer for the time when they would be required for analyses associating cluster formation with these variables. It is important to emphasize, of course, that the clustering is based solely on household expenditures and not on the exogenous variables just mentioned.

On the 590th pass and on the 629th pass (recall, there are 896 passes) seven clusters appeared, each with 15 or more households. These clustering configurations serve as the basis for our running analysis.

Before examining the degree of homogeneity in each cluster, it is desirable to see if the clusters have some association with the exogenous variables. Tables 9.1 and 9.2 give breakdowns for each of the seven clusters arising on the 590th pass and for the resulting universe. Without going into too much detail, since discussion appears in Suranyi-Unger (1977), the brief summary given in Table 9.2 indicates that the clusters each contain a different distribution of consumer characteristics; a glance at the table easily demonstrates this. Consider cluster number 2 and cluster number 5. Each contains 28 households, yet the ratios over age, family size, income and occupation vary markedly and as much as by two to two and a half.

Table 9.1
Clusters obtained in the 590th scan

Cluster number	Cluster size	Composition by consumer characteristics								
		*Age of family head		**Family size		†Family income		Occupation of family head		
		Low	High	Small	Large	Low	High	Professional	Sales and technical	Blue collar
1	55	21	34	27	28	34	21	10	10	35
2	28	15	13	10	18	23	5	3	3	22
3	73	27	46	39	34	36	37	15	16	42
4	18	9	9	7	11	8	10	3	4	11
5	28	10	18	15	13	20	8	4	5	19
6	39	13	26	22	17	29	10	5	5	29
7	15	4	11	6	9	6	9	4	3	8

*The age characteristic was defined with respect to 'high' or 'low' age of the head of the household. The dividing line here was set at the 40-year level.
**'Small' family size was defined as units consisting of one to three persons, 'large' as those numbering four or more.
†The income category was divided into classes of 'high', with an annual earning level of $7000 and above and 'low' earning under $7000 per year.

Table 9.2

Clusters of the 590th scan: ratios of consumer characteristics within and outside of the clusters

Cluster number	*Age ratio (low to high)	**Family size ratio (small to large)	†Income ratio (low to high)	Occupation ratios (Professional to sales-technical to blue collar)
1	1:1·61	1:1·03	1:0·62	1:0:1:3·50
2	1:0·87	1:1·80	1:1·03	1:1:7·33
3	1:1·70	1:0·87	1:1·03	1:1·06:2·80
4	1:1·00	1:1·57	1:1·25	1:1·33:3·66
5	1:1·80	1:0·87	1:0·40	1:1·25:4·75
6	1:2·00	1:0·77	1:0·34	1:1·00:5·80
7	1:2·75	1:1·50	1:1·50	1:1·04:3·77
Non-clustered portion of data universe	1:1·74	1:1·40	1:0·72	1:0·93:3·47

*The age characteristic was defined with respect to 'high' or 'low' age of the head of the household. The dividing line here was set at the 40-year level.

**'Small' family size was defined as units consisting of one to three persons, 'large' as those numbering four or more.

†The income category was divided into classes of 'high', with an annual earning level of $7000 and above and 'low' earning under $7000 per year.

With this information at hand, namely, that there is some homogeneity in household expenditures in each cluster and that, moreover, the clustering relates to the exogenous variables such as occupation, income, family size, etc., we now attempt to measure this. We borrow an index from hypothesis testing in multivariate analysis and employ it as a descriptive device to measure cluster homogeneity. This index is the ratio of the determinants (generalized variances) of two covariance matrices; the numerator is the determinant for the elements in the cluster as found while the denominator is the determinant of all the elements of the universe. The determinantal value is a one-dimensional summary (scalar) of the variability inherent in the system (cluster or universe). Thus, a ratio index of zero would indicate complete homogeneity of spending behavior, that is, each household within the cluster would allocate its income over 12 categories in an identical manner. An index of unity would produce no meaningful information because it would show that consumer behavior within the group is the same as over the whole population.

Each covariance matrix is 12 × 12 but, because the information vector of proportions sums up to unity in each case, each matrix has 11 eigenvalues. In order to obtain a numerical value for the ratio index, one would

Table 9.3
Ratio of the product of 9 largest eigenvalues of the determinant of the similarity matrix of cluster versus 896 observations used to form clusters

Cluster	Size	Ratio	
	First set of clusters		
1	55	0·0301	
2	28	0·0254	
3	73	0·3809	
4	18	0·00002	
5	28	0·0065	
6	39	0·0727	
7	15	0·0095	
Denominator = $2{\cdot}3237 \times 10^{13}$			
	Second set of clusters		
Cluster	Size	Ratio	
1	195	0·1561	
2	36	0·0003	
3	16	0·0035	
4	18	0·00002	(same as 4 above)
5	15	0·0095	(same as 7 above)
6	28	0·0065	(same as 5 above)
7	15	0·00011	

normally use the determinantal value of the similarity matrix (which is equivalent to the product of the 11 eigenvalues) in the numerator and in the denominator. However, since very small eigenvalues can throw matters off, each case requires special analysis. In Table 9.3, the decision was made to employ the product of the nine largest eigenvalues of the similarity matrix in the numerator and in the denominator. In the calculations underlying Table 9.3, the denominator was always the similarity matrix for the 896 households that were sampled from the 9476 urban households. The numerator reflects the cluster under consideration.

In short, Table 9.3 shows how each cluster differs in the degree of standardization (i.e., similarity of expenditure patterns) from the total data sample used in the clustering exercise.

Table 9.3 shows the standardization ratios for both sets of clusters, the first emerging from scan 590 and the second from scan 629. As is evident, each cluster shows a considerably higher degree of standardization of expenditure patterns than is found within the total data sample of 896 households.

Table 9.4

Ratios of determinants (11 dimensions) for demographic subgroups versus entire population (9476 observations)

	Size	Ratio
Age		
0-24 years	494	0·260
25-34 years	1847	0·045
35-49 years	3117	0·131
50-64 years	2396	1·657
65 + years	1622	1·691
Job type		
Professional	1784	0·092
Clerical/sales	1732	0·258
Blue collar	5960	1·599
Family size		
1 or 2 people	4372	7·755
3 or 4 people	3291	0·047
5 + people	1813	0·009
Income		
Under $4000	2614	12·481
$4000-$7000	3087	0·140
$7000-$10 000	2157	0·049
$10 000-$15 000	1185	0·038
Over $15 000	433	0·245

Table 9.5
Income-job type subgroups: ratios of determinants (11 dimensions) for demographic subgroups versus entire population (9476 observations)

Income	Job type	Size	Ratio
Under $4000	Professional	81	6·721
	Clerical/sales	306	2·655
	Blue collar	2227	10·554
$4000-$7000	Professional	385	0·040
	Clerical/sales	687	0·099
	Blue collar	2015	0·122
$7000-$10 000	Professional	576	0·024
	Clerical/sales	436	0·019
	Blue collar	1145	0·061
$10 000-$15 000	Professional	480	0·014
	Clerical/sales	216	0·031
	Blue collar	487	0·048
Over $15 000	Professional	262	0·074
	Clerical/sales	87	0·042
	Blue collar	84	0·651

The clustering thus suggests to the investigator that consumers characteristics are associated with consumer expenditures and the standardization index gives a picture of the homogeneity of spending behavior as a function of the endogenous variables. A next step for the investigator is to pursue these leads over the whole population of households. This was accomplished to give the results in Tables 9.4 to 9.7. In the exposition that follows we discuss the computations and provide an analysis.

Table 9.6
Job type-family size subgroups: ratio of determinants (11 dimensions) for demographic subgroups versus entire population (9476 observations)

Job type	Family size	Size	Ratio
Professional	1 or 2 people	620	1·2763
	3 or 4 people	752	0·0113
	5+ people	412	0·0012
Clerical/sales	1 or 2 people	759	2·0984
	3 or 4 people	621	0·0226
	5+ people	352	0·0024
Blue collar	1 or 2 people	2993	7·9694
	3 or 4 people	1918	0·0606
	5+ people	1049	0·0090

Table 9.7
Job type-age subgroups: ratio of determinants (11 dimensions) for demographic subgroups versus entire population (9476 observations)

Job type	Age	Size	Ratio
Professional	0-24 years	94	0·0037
	25-34 years	472	0·0052
	35-49 years	740	0·0230
	50-64 years	414	0·4403
	65 + years	64	2·5442
Clerical/sales	0-24 years	96	0·0792
	25-34 years	376	0·0725
	35-49 years	659	0·0252
	50-64 years	482	0·5582
	65 + years	119	0·0681
Blue collar	0-24 years	304	0·5077
	25-34 years	999	0·0349
	35-49 years	1718	0·2337
	50-64 years	1500	1·7522
	65 + years	1439	1·0090

It should be emphasized that in the calculations leading to the results in Tables 9.4 to 9.7 the problem of small eigenvalues did not arise, hence it was appropriate to use the determinantal values in arriving at the ratios. Again, an index of zero would mean complete homogeneity in spending patterns; an index of one simply denotes the degree of standardization in the data universe of 9476 households. Hence, any index under one signifies a higher degree of standardization and any value in excess of one a higher degree of heterogeneity (i.e. 'chaos') than is found in the data universe.

The standardization ratios were calculated for various combinations of the four consumer characteristics under investigation: occupation, income, age of the household head and size of the family.

Beginning with Table 9.4, which gives the standardization ratios within the various consumer characteristic groups, it becomes immediately obvious that a very high degree of standardization is found in the income ($7000 to $15 000 per annum) group and in the professional occupational group. As might be expected, these two groups embody considerable overlap because, of the 1784 professionals in the universe, 1056 or 59% belong to this income class, whereas the 1629 blue-collar members of this income class comprise only 27% of the blue-collar component of the entire population. As noted above, this blue-collar group forms a behavior group of its own, possibly similar in spending behavior to the professional class in the same income range. Another observation that should not be neglected is that the ratios for family size in categories '3 or 4' and '5 or over' are among the smallest and suggest very homogeneous spending behavior in these groups.

The next noteworthy feature of the income-occupation relationship is revealed by Table 9.5, where it is evident that a high degree of standardization is found in the professional classes and that in every income bracket examined, the degree of professional standardization consistently outweighs the degree of standardization in the blue-collar component of the income class. According to Table 9.6, professionals show more standardization than blue-collar workers in every family-size group depicted, and Table 9.7 shows that with the exception of the over 65-year age group (consisting of only 64 observations compared to 1439 in the blue-collar group) professionals are consistently more standardized in their lifestyle than the blue-collar populations in every one of the age groups displayed.

Diversified (or chaotic) spending behavior is found, first and foremost, in the low-income classes and in the blue-collar worker group with small familes (including single consumers). Regardless of the degree of standardization or diversification, however, in all computations the white-collar/professional groups consistently show a higher degree of standardization than their blue-collar counterparts within the same demographic group (as measured by criteria other than occupation). This finding is shown in Tables 9.5 to 9.7. It is also interesting to observe in an examination of spending for the single commodities within the expenditure matrix used throughout the research, that in all income ranges over $4000, the white-collar/professional groups consistently outspend the blue-collar classes for housing, education and reading, whereas the blue-collar groups outspend on food and alcohol.

A most interesting phenomenon of these data is the higher homogeneities (smaller ratios) found for almost all seven clusters in comparison to those for the various classes corresponding to the individual 'exogenous' variables. This most certainly emphasizes the value of clustering multivariate data profiles over univariate or bivariate (cross-classification) descriptions.

Much more can be said about the economic implications of these results and the reader is referred to Suranyi-Unger (1977). For the purposes of this chapter we have demonstrated an application of the results of our inquiries on clustering procedures plus an extension on how homogeneity or standardization within clusters can be measured.

References

Abrams, R., Taylor, M. A. and Gastanaga, P. (1974). Manic-depressive illness and paranoid schizophrenia. *Archives of General Psychiatry,* **31,** 640-642.

American Psychiatric Association (1968). "Diagnostic and Statistical Manual of Mental Disorders", 2nd edn. Washington, D.C., A.P.A.

American Psychiatric Association (1980). "Diagnostic and Statistical Manual of Mental Disorders", 3rd edn (DSM-III). Washington, D.C., A.P.A.

Anderberg, M. R. (1973). "Cluster Analysis for Applications". New York, Academic Press.

Baker, F. B. (1974). Stability of two hierarchical grouping techniques. Case I: Sensitivity to data errors. *Journal of the American Statistical Association,* **69,** 440-445.

Ball, G. H. (1970). "Classification analysis". Menlo Park, California, Stanford Research Institute.

Bali, G. H. and Hall, D. J. (1965). "ISODATA, A Novel Method of Data Analysis and Pattern Classification". (AD 699616) Menlo Park, California, Stanford Research Institute.

Ball, G. H. and Hall, D. J. (1967). A clustering technique for summarizing multivariate data. *Behavioral Science,* **12,** 153-155.

Bartko, J. J., Strauss, J. S. and Carpenter, W. T. (1971). An evaluation of taxometric techniques for psychiatric data. *Classification Society Bulletin,* **2,** 2-28.

Bartlett, H. H. (1940). *Bulletin Torrey Botanical Club,* **67,** 349.

Blashfield, R. K. and Draguns, J. G. (1976). Evaluative criteria for psychiatric classification. *Journal of Abnormal Psychology,* **85,** 140-150.

Boyce, A. J. (1969). Mapping diversity: A comparative study of some numerical methods. In: A. J. Cole (Ed.) "Numerical Taxonomy. Proceedings of the Colloquium in Numerical Taxonomy held in the University of St. Andrews, September 1968". London, Academic Press.

Camper, P. (1979). "Dissertation Physique sur les Différences Réelles que Présentent les Traits du Visage chez les Hommes de Différents Pays et de Différents Ages". Utrecht.

Carlson, K. A. (1972). A method for identifying homogeneous classes. *Multivariate Behavioral Research,* **7,** 483-488.

Carmichael, J. W., George, J. A. and Julius, R. S. (1968). Finding natural clusters. *Systematic Zoology,* **17,** 144-150.

Carmichael, J. W. and Sneath, P. H. A. (1969). Taxometric maps. *Systematic Zoology,* **18,** 402-415.
Carpenter, W. T., Bartko, J. J., Carpenter, C. L. and Strauss, J. S. (1967). Another view of schizophrenic subtypes. *Archives of General Psychiatry,* **33,** 508-516.
Cattell, R. B. and Coulter, M. A. (1966). Principles of behavioral taxonomy and the mathematical basis of the taxonome computer program. *British Journal of Mathematical and Statistical Psychology,* **19,** 237-269.
Chernoff, H. (1973). The use of faces to represent points in *k*-dimensional space graphically. *Journal of the American Statistical Association,* **68,** 361-368.
Cormack, R. M. (1971). A review of classification. *Journal of the Royal Statistical Society A,* **134,** 321-367.
Cramér, H. (1946). "The Elements of Probability Theory and some of its Applications". New York, John Wiley and Sons.
Cronbach, L. J. (1975) Beyond the two disciplines of scientific psychology. *American Psychologist,* **30,** 116-127.
Cumming, J. and Cumming, E. (1957). Social equilibrium and social change in a large mental hospital. In: M. Greenblatt, D. Levinson and R. Williams (Eds) "The Patient and the Mental Hospital". Glencoe, Ill., Free Press.
Cunningham, K. M. and Ogilvie, J. C. (1972). Evaluation of hierarchical grouping techniques—A preliminary study. *Computer Journal,* **15,** 209-215.
Czekanowski, J. (1911). Objektive Kriterien in der Ethnologie. *Korrespondez-Blatt der Deutschen Gesellschaft für Anthropologie, Ethnologie und Urgeschichte,* **42,** 1-5.
Degerman, R. (1970). Multidimensional analysis of complex structure mixture of class and quantitative variation. *Psychometrika,* **35,** 475-491.
Driver, H. E. (1965). Survey of numerical classification in anthropology. In: D. H. Hymes (Ed.) "Symposium on the Use of Computers in Anthropology". The Hague, Mouton.
Driver, H. E. and Schuessler, K. F. (1957). Factor analysis of ethnographic data. *American Anthropologist,* **59,** 655-663.
Everitt, B. (1974). "Cluster Analysis". New York, John Wiley and Sons.
Fabrega, H., Jr (1976). The biological significance of taxonomies of disease. *Journal of Theoretical Biology,* **63,** 191-216.
Feinstein, A. R. (1967). "Clinical judgment". Huntington, N.Y., Robert E. Krieger.
Fisher, R. A. (1936). The use of multiple measurements in taxonomic problems. *Annals of Eugenics,* **7,** 179-188.
Forgy, E. W. (1965). Cluster analysis of multivariate data: efficiency versus interability of classifications. *Biometrics,* **21,** 768.
Fortier, J. J. and Solomon, H. (1966). Clustering procedures. In: P. R. Krishnaiah (Ed.) "Multivariate Analysis". New York, Academic Press.
Friedman, H. P. and Rubin, J. (1967). On some invariant criteria for grouping data. *Journal of the American Statistical Association,* **62,** 1159-1178.
Good, I. J. (1965). Categorization of classification. In: "Mathematics and Computer Science in Biology and Medicine". London, HMSO.
Green, P. E. and Rao, V. R. (1969). A note on proximity measures and cluster analysis. *Journal of Marketing Research,* **6,** 359-364.
Hartigan, J. A. (1972). Direct clustering of a data matrix. *Journal of the American Statistical Association,* **67,** 123-129.
Hartigan, J. A. (1975). "Clustering Algorithms". New York, John Wiley and Sons.
Hempel, C. (1965). "Aspects of Scientific Explanation". New York, The Free Press.

Holzinger, K. J. and Harman, H. H. (1941). "Factor analysis". Chicago, University of Chicago Press.

Hubert, L. (1974). Approximate evaluation techniques for the single-link and complete-link hierarchical clustering procedures. *Journal of the American Statistical Association,* **69,** 698-704.

Jancey, R. C. (1966). Multidimensional group analysis. *Australian Journal of Botany,* **14,** 127-130.

Jardine, J. and Sibson, R. (1968). The construction of hierarchic and non-hierarchic classifications. *Computer Journal,* **11,** 117-184.

Johnson, S. C. (1967). Hierarchical clustering schemes. *Psychometrika,* **32,** 241-254.

Jones, M. (1953). "The Therapeutic Community: A New Treatment Method in Psychiatry". New York, Basic Books.

Kahl, J. A. and Davis, J. A. (1955). A comparison of indexes of socio-economic status. *American Sociological Review,* **20,** 317-325.

Kay, P. (1971). "Explorations in Mathematical Anthropology". Cambridge, Mass., The MIT Press.

Kendall, M. G. (1966). Discrimination and classification. In: P. R. Krishnaiah (Ed.) "Multivariate Analysis". New York, Academic Press.

Kidder, A. V. (1915). Pottery of the Pajarito Plateau and of some adjacent regions in New Mexico. *American Anthropological Association,* Memoir **2,** 407-462.

King, B. F. (1966). Market and industry factors in stock price behavior. *Journal of Business,* **39,** 139-190.

King, B. F. (1967). Step-wise clustering procedures. *Journal of the American Statistical Association,* **62,** 86-101.

Kroeber, A. L. (1916). Zuni potsherds. *American Museum of Natural History, Anthropological Papers,* **18,** 1-38.

Kroeber, A. L. and Dixon, R. B. (1903). Native languages of California. *American Anthropologist,* **5,** 1-26.

Kruskal, J. B. (1964a). Multidimensional scaling by optimizing goodness of fit to a non-metric hypothesis. *Psychometrika,* **29,** 1-27.

Kruskal, J. B. (1964b). Non-metric multidimensional scaling: A numerical method. *Psychometrika,* **29,** 115-129.

Lambert, J. M. and Williams, W. T. (1962). Multivariate methods in plant ecology. IV. Nodal analysis. *Journal of Ecology,* **50,** 775-802.

Ling, R. F. (1972). On the theory and construction of *k*-clusters. *Computer Journal,* **15,** 326-332.

Ling, R. F. (1973). A probability theory of cluster analysis. *Journal of the American Statistical Association,* **68,** 159-164.

Linnaeus, C. (1737). "Genera Plantarum". Uppsala.

Lorr, M. and Radhakrishnan, B. K. (1967). A comparison of two methods of cluster analysis. *Educational and Psychological Measurement,* **27,** 47-53.

MacQueen, J. B. (1967). Some methods for classification and analysis of multivariate observations. *Proceedings of the Fifth Berkeley Symposium on Mathematical Statistics and Probability,* **1,** 281-297.

Mezzich, J. E. (1979). Patterns and issues in multiaxial psychiatric diagnosis. *Psychological Medicine,* **9,** 125-137.

Mezzich, J. E. (1980). Multiaxial diagnostic systems in psychiatry. In: Kaplan, H. I., Freedman, A. M. and Sadock, B. J. (Eds) "Comprehensive Textbook of Psychiatry", 3rd edn. Baltimore, Williams and Wilkins.

Mezzich, J. E. and Worthington, R. D. L. (1978). A comparison of graphical representations of multi-dimensional psychiatric diagnostic data. In: Peter C. C. Wang (Ed.) "Graphical Representation of Multivariate Data". London, Academic Press.

Moos, R. H. (1974a). "Ward Atmosphere Scale Manual". Palo Alto, California, Consulting Psychologists Press.

Moos, R. H. (1974b). "Evaluating Treatment Environments: A Social Ecological Approach". New York, John Wiley and Sons.

Moos, R. H., Shelton, R. and Petty, C. (1973). Perceived ward climate and treatment outcome. *Journal of Abnormal Psychology,* **82,** 291-298.

Overall, J. E. and Gorham, D. R. (1962). The Brief Psychiatric Rating Scale. *Psychological Reports,* **10,** 799-812.

Overall, J. E. and Klett, C. J. (1972). "Applied Multivariate Analysis". New York, McGraw-Hill.

Panzetta, A. F. (1974). Toward a scientific psychiatric nosology. *Archives of General Psychiatry,* **30,** 154-161.

Piaget, J. (1973). "The Psychology of Intelligence". New Jersey, Littlefield, Adams and Co.

Piazza, A., Sgaramella-Zonta, L., Gluckman, P. and Cavalli-Sforza, L. L. (1975). The Fifth Histocompatibility Workshop: Gene frequency data—A phylogenetic analysis. *Tissue Antigens,* **5,** 445-463.

Price, R. H. and Moos, R. H. (in press). Toward a taxonomy of inpatient treatment environments. *Journal of Abnormal Psychology,* **82.**

Quetelet, A. (1835). "Essai de Phipegne Sociale". Paris.

Randolph, L. F. (1934). Chromosome numbers in native American and introduced species and cultivated varieties of Iris. *Bulletin of the American Iris Society,* **52,** 61-66.

Raven, P. H., Berlin, B. and Breedlove, D. E. (1971). The origins of taxonomy. *Science,* **174,** 1210-1213.

Rogers, G. and Linden, J. D. (1973). Use of multiple discriminant function analysis in the evaluation of three multivariate grouping techniques. *Educational and Psychological Measurement,* **33,** 787-802.

Rosenthal, D and Ketty, S. (Eds) (1968). "The Transmission of Schizophrenia". Oxford, Pergamon Press.

Rubin, J. and Friedman, H. P. (1967). "A Cluster Analysis and Taxonomy System for Grouping and Classifying Data-Computer Program". New York, IBM Corp.

Rutter, M., Schaffer, D. and Shepherd, M. (1975). "A Multiaxial Classification of Child Psychiatric Disorders". Geneva: World Health Organization.

Shepard, R. N. (1962a). The analysis of proximities: Multidimensional scaling with an unknown distance function. I. *Psychometrika,* **27,** 125-139.

Shepard, R. N. (1962b). The analysis of proximities: Multidimensional scaling with an unknown distance function. II. *Psychometrika,* **27,** 219-246.

Shepard, R. N., Romney, A. K. and Nerlove, S. B. (1972). "Multidimensional Scaling. Theory and Application in the Behavioral Sciences". Vol. 1: Theory. New York, Seminar Press.

Skinner, H. A., Reed, P. L. and Jackson, D. N. (1976). Toward the objective diagnosis of psychopathology: Generalizability of modal personality profiles. *Journal of Consulting and Clinical Psychology,* **44,** 111-117.

Sneath, P. H. A. (1966). A comparison of different clustering methods as applied to randomly spaced points. *Classification Society Bulletin,* **1,** 2-18.

Sneath, P. H. A. and Sokal, R. R. (1973). "Numerical Taxonomy. The Principles and Practice of Numerical Classification". San Francisco, Freeman.

Sokal, R. R. and Rohlf, F. J. (1962). The comparison of dendrograms by objective methods. *Taxon,* **11,** 33—40.

Solomon, H. (1971). "Numerical Taxonomy" Mathematics in the Archaeological and Historical Sciences, 62-81. Edinburgh, Edinburgh University Press.

Spearman, C. (1927). "The Abilities of Man, their Nature and Measurement". New York, Macmillan.

Spence, N. A. and Taylor, P. J. (1970). Quantitative methods in regional taxonomy. *Progress in Geography,* **2,** 1-64.

Spier, L. (1917). An outline for a chronology of Zuni ruins. *Anthropological Papers of the American Museum of Natural History,* **18,** 207-331.

Stephenson, W. (1935). Correlating persons instead of tests. *Character and Personality,* **4,** 17-24.

Strauss, J. S. (1973). Diagnostic models and the nature of psychiatric disorder. *Archives of General Psychiatry,* **29,** 445-449.

Strauss, J. S. (1975). A comprehensive approach to psychiatric diagnosis. *American Journal of Psychiatry,* **132,** 1193-1197.

Strauss, J. S., Bartko, J. J. and Carpenter, W. T. (1973). The use of clustering techniques for the classification of psychiatric patients. *British Journal of Psychiatry,* **122,** 531-540.

Strong, W. D. (1925). Uhle pottery collections from Ancon. *University of California Publications in American Archaeology and Ethnology,* **21,** 135-190.

Suranyi-Unger, T., Jr (1977). "Identification of Standard Classes in the United States". Washington, D.C., National Science Foundation, NSF/RA77-0205.

Temkin, O. (1965). The history of classification in the medical sciences. In: M. Katz, J. Cole and W. Barton (Eds). "Classification in Psychiatry and Psychopathology". Washington, D. C., Government Printing Office.

Thurstone, L. L. (1938). "Primary Mental Abilities". Psychometric Monographs No. 1. Chicago: University of Chicago Press.

Thurstone, L. L. (1947). "Multiple Factor Analysis". Chicago: University of Chicago Press.

Tryon, R. C. (1939). "Cluster Analysis". Ann Arbor, Edward Bros.

Ward, J. H. (1963). Hierarchical grouping to optimize an objective function. *Journal of the American Statistical Association,* **58,** 236-244.

Wherry, R. J., Sr and Wherry, R. J., Jr (1971). Wherry-Wherry hierarchical factor analysis. In: R. J. Wherry and J. Olivero (Eds) "Computer Programs for Psychology". Columbus, Ohio, Department of Psychology, The Ohio State University.

Winokur, G., Clayton, P. J. and Reich, T. (1969). "Manic-depressive Illness". St. Louis, C. V. Mosby.

Wishart, D. (1969a). Mode analysis. In: A. J. Cole (Ed) "Numerical Taxonomy". New York, Academic Press.

Wishart, D. (1969b). Numerical classification method for deriving natural classes. *Nature,* **221,** 97-98.

Wolfe, J. H. (1965). "A Computer Program for the Maximum Likelihood Analysis of Types" (Technical Bulletin, 65-15). San Diego, California, US Naval Personnel Research Activity.

Wolfe, J. H. (1967). "NORMIX Computational Methods for Estimating the Parameters of Multivariate Normal Mixtures of Distributions" (Research Memorandum, SRM 68-2). San Diego, California, US Naval Personnel Research Activity.

Wolfe, J. H. (1970). Pattern clustering by multivariate mixture analysis. *Multivariate Behavioral Research,* **5,** 329-350.

Wolfe, J. H. (1971). "NORMIX 360 Computer Program" (R. M. SRM 72-4). San Diego, California, US Naval Personnel and Training Research Laboratory.

World Health Organization (1978). "International Classification of Diseases, 9th Revision (ICD-9)". Geneva, WHO.

Zubin, J. (1938). Socio-biological types and methods for their isolation. *Psychiatry,* **1,** 237-247.

Index

QUANTITATIVE STUDIES IN SOCIAL RELATIONS

Consulting Editor: Peter H. Rossi

UNIVERSITY OF MASSACHUSETTS
AMHERST, MASSACHUSETTS

Peter H. Rossi and Walter Williams (Eds.), EVALUATING SOCIAL PROGRAMS: ***Theory, Practice, and Politics***

Roger N. Shepard, A. Kimball Romney, and Sara Beth Nerlove (Eds.), MULTIDIMENSIONAL SCALING: ***Theory and Applications in the Behavioral Sciences,*** Volume I – Theory; Volume II – Applications

Robert L. Crain and Carol S. Weisman, DISCRIMINATION, PERSONALITY, AND ACHIEVEMENT: ***A Survey of Northern Blacks***

Douglas T. Hall and Benjamin Schneider, ORGANIZATIONAL CLIMATES AND CAREERS: ***The Work Lives of Priests***

Kent S. Miller and Ralph Mason Dreger (Eds.), COMPARATIVE STUDIES OF BLACKS AND WHITES IN THE UNITED STATES

Robert B. Tapp, RELIGION AMONG THE UNITARIAN UNIVERSALISTS: ***Converts in the Stepfathers' House***

Arthur S. Goldberger and Otis Dudley Duncan (Eds.), STRUCTURAL EQUATION MODELS IN THE SOCIAL SCIENCES

Henry W. Riecken and Robert F. Boruch (Eds.), SOCIAL EXPERIMENTATION: ***A Method for Planning and Evaluating Social Intervention***

N. J. Demerath, III, Otto Larsen, and Karl F. Schuessler (Eds.), SOCIAL POLICY AND SOCIOLOGY

H. M. Blalock, A. Aganbegian, F. M. Borodkin, Raymond Boudon, and Vittorio Capecchi (Eds.), QUANTITATIVE SOCIOLOGY: ***International Perspectives on Mathematical and Statistical Modeling***

Carl A. Bennett and Arthur A. Lumsdaine (Eds.), EVALUATION AND EXPERIMENT: ***Some Critical Issues in Assessing Social Programs***

Michael D. Ornstein, ENTRY INTO THE AMERICAN LABOR FORCE

Seymour Sudman, **APPLIED SAMPLING**

James D. Wright, **THE DISSENT OF THE GOVERNED:** ***Alienation and Democracy in America***

Roland J. Liebert, **DISINTEGRATION AND POLITICAL ACTION:** ***The Changing Functions of City Governments in America***

Walter Williams and Richard F. Elmore, **SOCIAL PROGRAM IMPLEMENTATION**

Edward O. Laumann and Franz U. Pappi, **NETWORKS OF COLLECTIVE ACTION:** ***A Perspective on Community Influence Systems***

Eric Hanushek and John Jackson, **STATISTICAL METHODS FOR SOCIAL SCIENTISTS**

Richard F. Curtis and Elton F. Jackson, **INEQUALITY IN AMERICAN COMMUNITIES**

Richard A. Berk, Harold Brackman, and Selma Lesser, **A MEASURE OF JUSTICE:** ***An Empirical Study of Changes in the California Penal Code, 1955–1971***

Samuel Leinhardt (Ed.), **SOCIAL NETWORKS:** ***A Developing Paradigm***

Donald J. Treiman, **OCCUPATIONAL PRESTIGE IN COMPARATIVE PERSPECTIVE**

Beverly Duncan and Otis Dudley Duncan, **SEX TYPING AND SOCIAL ROLES:** ***A Research Report***

N. Krishnan Namboodiri (Ed.), **SURVEY SAMPLING AND MEASUREMENT**

Robert M. Groves and Robert L. Kahn, **SURVEYS BY TELEPHONE:** ***A National Comparison With Personal Interviews***

Peter H. Rossi, Richard A. Berk, and Kenneth J. Lenihan, **MONEY, WORK, AND CRIME:** ***Experimental Evidence***

Walter Williams, **GOVERNMENT BY AGENCY:** ***Lessons From The Social Program Grants-In-Aid Experience***

In Preparation

Juan E. Mezzich and Herbert Solomon, **TAXONOMY AND BEHAVIORAL SCIENCE**

Zev Klein and Yohanan Eshel, **INTEGRATING JERUSALEM SCHOOLS**

Philip K. Robins, Robert G. Spiegelman, Samuel Weiner, and Joseph G. Bell (Eds.), **A GUARANTEED ANNUAL INCOME:** ***Evidence From a Social Experiment***